UNIVERSITY CONSORTIUM FOR
GEOGRAPHIC INFORMATION SCIENCE

Computation and Visualization for Understanding Dynamics in Geographic Domains

A Research Agenda

T0262661

UNIVERSITY CONSORTIUM FOR
GEOGRAPHIC INFORMATION SCIENCE

Computation and Visualization for Understanding Dynamics in Geographic Domains

A Research Agenda

May Yuan

Kathleen Stewart Hornsby

CRC Press
Taylor & Francis Group
Boca Raton London New York

CRC Press is an imprint of the
Taylor & Francis Group, an **informa** business

CRC Press
Taylor & Francis Group
6000 Broken Sound Parkway NW, Suite 300
Boca Raton, FL 33487-2742

© 2008 by Taylor & Francis Group, LLC
CRC Press is an imprint of Taylor & Francis Group, an Informa business

Library of Congress Cataloging-in-Publication Data

Yuan, May.
    Computation and visualization for the understanding of dynamics in geographic domains : a research agenda / May Yuan and Kathleen S. Hornsby.
        p. cm.
    Includes bibliographical references and index.
    ISBN 978-1-4200-6032-4 (alk. paper)
    1. Geographic information systems. 2. Geodatabases. 3. Geography--Computer network resources. I. Hornsby, Kathleen. II. Title.

G70.212.Y83 2008
910.01'1--dc22   2007037310

Visit the Taylor & Francis Web site at
http://www.taylorandfrancis.com

and the CRC Press Web site at
http://www.crcpress.com

CONTENTS

CHAPTER 1

INTRODUCTION

The world is ever changing, and a comprehensive understanding of the world will not be achieved without theoretical and methodological advances to decode complex dynamics in human and environmental systems. Dynamics refers to happenings in the world that involve change or movement. Oil spills, tornadoes, and protest marches are all examples of geographic dynamics. In addition to scientific inquiries, understanding these dynamics is essential across all levels of human and environmental dimensions—from individual daily activities to world peacekeeping operations, from local land use decisions to global climate change.

To investigate research developments and the challenges to visualization and computation of geographic dynamics, the University Consortium for Geographic Information Science (UCGIS), with sponsorship from the Disruptive Technology Office (DTO) on behalf of the U.S. Intelligence Community (IC), hosted a workshop on Computation and Visualization for the Understanding of Dynamics in Geographic Domains. In short, the Dynamics Workshop was the second workshop organized through a UCGIS–IC joint effort to tackle research issues at the intersection of geographic information science and intelligence analysis. The precursor to the Dynamics workshop was the 2003 UCGIS workshop on Geospatial Visualization and Knowledge Discovery[1] with a focus on research needs linking geovisualization with knowledge discovery for the detection and analysis of unexpected events in space and time. The 2006 Dynamics workshop focused further on spatiotemporal topics, highlighting geographic dynamics and the visual and computational methods that can improve their understanding, particularly for intelligence analysis. The co-organizers for

---

[1] http://www.ucgis.org/Visualization

this workshop were May Yuan (University of Oklahoma) and Kathleen Stewart Hornsby[2] (University of Maine).

I. WORKSHOP

A call for position papers was distributed to all UCGIS member institutions, including seventy-one universities, four professional organizations, five corporate affiliates, one U.S. government agency, and two international associations. All submitted position papers were reviewed by members of the workshop steering committee (Table 1.1), and fourteen geographic information scientists were invited based on their position papers. Four leading scholars in related fields were invited to deliver plenary talks. Additional UCGIS participants, including representatives from industry, brought the number of non-government attendees to twenty-five. Rounding out the participants were a further twenty-five representatives from the US intelligence community. The workshop was held October 16 to 18, 2006 at the Maritime Institute of Technology and Graduate Studies, Linthicum Heights, Maryland.

Table 1.1: Dynamics Workshop Steering Committee

Committee Members	Institutions
Jiawei Han	University of Illinois at Urbana–Champaign
Kathleen Stewart Hornsby	University of Maine
Nina Lam	Louisiana State University
Alan MacEachren	The Pennsylvania State University
Harvey Miller	University of Utah
May Yuan	University of Oklahoma

The workshop offered a multidisciplinary and multi-agency perspective on the latest research work relating to visualization and computation of geographic dynamics. UCGIS participants included academic and industrial researchers, representing disciplines from geography, spatial information science and engineering, history, epidemiology, environmental

---

[2] Kathleen Stewart Hornsby joined the University of Iowa in 2007.

science, computer science, and cognitive science. One workshop participant represented Geomatics for Informed Decisions (GEOIDE) in Canada, a similar organization to the UCGIS. Government participants were from the US intelligence community, United States Geological Surveys (USGS), and the National Institutes of Health (NIH). Table 1.2 lists all UCGIS workshop participants.

Table 1.2: Workshop Participants from UCGIS Institutions and Affiliates

Participants	Institutions
Sean Ahearn	Hunter College, City University of New York
Ola Ahlqvist	The Ohio State University
Michael Goodchild	University of California–Santa Barbara
Elizabeth Groff	University of Maryland
Dustin Howard	University of Oklahoma
Steve Kopp	Environmental Systems and Research Institute
Mei Po Kwan	The Ohio State University
Nina Lam	Louisiana State University
Marguerite Madden	University of Georgia
Jaymie Meliker	University of Michigan/BioMedware, Inc.
Jack Owens	Idaho State University
Donna Peuquet	The Pennsylvania State University
Steven Prager	University of Wyoming
Chris Renschler	State University of New York at Buffalo
William Ribarsky	University of North Carolina–Charlotte
Darren Scott	McMaster University
Cyrus Shahabi	University of Southern California
Narushige Shiode	State University of New York at Buffalo
Kathleen Stewart Hornsby	University of Maine
Paul Torrens	Arizona State University
Colin Ware	University of New Hampshire
John Wilson	University of Southern California
Ningchuan Xiao	The Ohio State University
May Yuan	University of Oklahoma

The workshop was organized into sessions of plenary presentations, scenario analysis, government panel discussions, academic research briefings, and breakout group discussions. In the plenary session, four leading scholars summarized the state of research developments and highlighted projects on visual analytics and computational models facilitating geographic dynamics understanding. Topics included: visual analysis of urban terrain dynamics, visual analysis of human activities, why do we keep turning time into space? and representation and computation of geographic dynamics. These presentations set the vision and challenges for subsequent workshop discussions. Following the plenary session, a problem-solving exercise provided workshop participants hands-on experience with the nature of intelligence reporting as well as thought exercises on evidence-based intelligence analysis. Participants worked in groups to generate plausible hypotheses about events in space and time and to construct defensible and persuasive arguments for certain hypotheses. Throughout the exercise, the participants discussed visual and computational methods and their effectiveness for spatiotemporal analysis and prediction of events.

A government panel of representatives from throughout the intelligence community outlined the needs and challenges of spatial and temporal analysis and modeling for their agencies. These needs and challenges can lead to an understanding of activities, events, and processes in geographic domains and produce actionable intelligence to achieve their missions. In response, UCGIS participants briefed the group on their academic research projects, covering visual and computational methods to address a wide range of geographic dynamics and application domains. The government panel and academic research briefings set the stage for vivid and inspiring group discussions. In the final session of the workshop, suggestions were made on how to sustain the momentum of these discussions as well as the determination to advance GIScience research on geographic dynamics. The interactions between UCGIS and government participants have built a consensus view of the challenges and possibilities to make the next big strike in advancing visualization and computation for the understanding of dynamics in geographic domains.

II.   REPORTING ON THE WORKSHOP

This book is an extended report of the workshop. It's goal is to synthesize key ideas and issues discussed during the workshop and expand upon the workshop discussions to provide a comprehensive overview of geographic dynamics and approaches to advancing the understanding through computation and visualization. A basic assumption of the workshop posits that understanding our world requires a thorough comprehension of dynamics in geographic domains at all levels. Our world knowledge is built upon our abilities to analyze spatiotemporal patterns, reason on spatiotemporal relationships, model spatiotemporal behaviors, and predict what comes next in space and time. In addition, information systems, such as geographic information systems, have lagged in their ability to support the modeling, visualization, and computation that are required to carry out analysis of geographic dynamics.

Dynamics in geographic domains cuts across a wide spectrum of spatiotemporal themes, ranging from individual travel activities to global environmental change. On one scale, dynamics reflects the movement patterns or the change to entities over time, where entities go in and out of existence, as well as the internal evolution of properties of entities; on another scale, dynamics is manifested through interactions among geographic systems, restructuring these systems, and developing new systems. For example, the interactions between human systems and environmental systems display dynamics that can drive changes to both systems. Hence, a thorough comprehension of dynamics in geographic domains challenges all aspects of information computation and visualization from the development of effective analytical, reasoning, and computational environments to understanding and modeling intricate relationships and interactions evolving in our ever-changing world.

The goal of this book is to establish a foundation to promote research in geographic dynamics and provide a springboard for the next big scientific and technology breakthrough in this research area. Chapters 1 through 5 give an overview of the state of research developments in geographic dynamics and how this research relates to intelligence analysis. In Chapter 2, we synthesize materials from the plenary speeches as well as additional research findings from the broader literature to address the nature and dimensions of dynamics in geographic domains. We also examine the fundamentals, mechanisms, and functions that define geographic dynamics. In Chapter 3, we examine approaches to spatiotemporal data handling and identify concepts and ideas in geographic dynamics that can add new dimensions for the detection of new features, contextualization of existing evidences, or other means to incorporate information about geographic dynamics in GIS databases. In Chapters 4 and 5, the visual and computational methods for geographic dynamics research are discussed also in relation to intelligence analysis. Materials are synthesized from presentations at the workshop with substantial supplements from additional literature reviewed by the authors.

Chapter 6 offers perspectives for future research. We fuse suggestions from workshop participants with literature reviews to propose research agenda and recommendations for future developments and collaboration. We have made every effort to cite and acknowledge key ideas contributed from individual workshop participants, except for summaries reported from group discussions However, we apologize for any oversights or omissions in our notes. While the workshop was sponsored by the intelligence community, hosted by the UCGIS, and supported by excellent presentations, briefings, and discussions by the workshop participants, responsibility for the final content of this book rests with the authors.

III.   ACKNOWLEDGMENTS

The Disruptive Technology Office provided funding support for the workshop. The University Consortium for Geographic Information Science administered the workshop and post-workshop activities. May Yuan's research is supported in part by the U.S. National Science Foundation through Collaborative Awards BCS-0416208. Kathleen Stewart Hornsby's research is supported in part by a grant from the National Geospatial-Intelligence Agency HM1582-05-1-2039. The authors would like to express their appreciation to Arnold Landvoigt and Jack Sanders for their support and encouragement, to Tai Soda, Susan Horwitz, and Catherine Ciacari for their efforts in coordinating the book publication, and to Nicholas Benson for his assistance on figures, tables, and references.

CHAPTER 2

THE NATURE OF GEOGRAPHIC DYNAMICS

- *What is geographic dynamics?*
- *How does geographic dynamics manifest itself?*
- *How does research in geographic dynamics benefit intelligence analysis?*

The word "dynamics" conveys ideas concerning change or movement where, for both, the concept of time is essential. A system (i.e., a set of interacting elements that comprise a unified whole) is characterized as dynamic when its parts are temporally dependent such that what happens to one part of the system triggers adjustments to some or all other parts. Events that act upon the system, activities that occur within the system, and processes that transform the system drive the dynamics of the system and result in change and movement as responses to these happenings and adjustments. In this chapter, we explore the nature of geographic dynamics in reality and determine the fundamental elements for understanding geographic dynamics. All terminologies used in this chapter refer to existence in geographic worlds. The following chapter will make the connections between the real world and how it is represented in information systems.

Many disciplines study dynamics in specific domains of interest, such as atmospheric dynamics or social dynamics, to understand the fundamental functions of relevant forces and their interrelationships in space and time. The scope of geographic dynamics is extensive; geographic dynamics encompasses all time-dependent aspects of physical and human systems on or near the surface of the Earth (i.e., the geographic domain). When delivering his plenary talk, Dr. Michael Goodchild pointed out that geographic dynamics spans the concerns of all disciplines that deal with change in geographic space: geography, meteorology, transportation, criminology, sociology, archaeology, and many more. He suggested that the role of GIScience research is to develop generic theories, tools, languages, and data models for studying geographic dynamics of all kinds. Such GIScience research requires understanding the full domain as well as the individual elements from which geographic dynamics can be understood. One challenge in GIScience research relates to its multidisciplinary nature, where disciplines may use the same term but adopt different concepts for this term. This chapter centers upon the concepts and terminologies used to communicate and characterize geographic dynamics by domain scientists, such as hydrologists and sociologists. The next chapter makes connections between terminologies adopted in domain sciences and information systems.

I.   DEFINITIONS

By developing generic theories, research on geographic dynamics addresses the mechanism by which geographic drivers interact with objects, people, phenomena, and places as well as the outcomes of the interactions. Geographic drivers refer to spatiotemporal activities, events, and processes, and change and movement are the outcomes of such interactions. Geographic dynamics are embedded in the work of these drivers and their outcomes. Spatial and temporal considerations are essential, and the study of geographic dynamics must be able to account for drivers, elements, mechanisms, and consequences in space and time. While activities, events, and processes are familiar terms to many readers, clear assertions of these terms in the context of geographic dynamics are necessary for further discussions in this book. The definitions in Table 2.1 expand upon entries in the Oxford English Dictionary to the context of geographic dynamics.

Table 2.1: Definitions

Term	Oxford English Dictionary	Geographic Dynamics
Activity	The state or quality of being active; exertion of energy; the degree to which a substance, esp. an enzyme, exhibits its characteristic property.	Action or doing taken by individuals (e.g., objects or humans) in space and time. The consequence of activities by an individual may generate movement

		of the individual or cause change to the individual's characteristics.
Event	The fact of anything happening; the occurrence of; an incident.	An occurrence of something with significance that drives noteworthy change at locations over time. The decision on "significance" and "noteworthy" is situational and problem-dependent.
Process	The fact of going on or being carried out, as an action, or a series of actions or events; progress, course.	A gradual transformation that transcends geographic properties, forms, and patterns over time. The determination of "gradualness" is scale-dependent.
Change	Substitution of one thing for another; succession of one thing in place of another; substitution of other conditions or circumstances, variety.	Substitution of properties in an object, at a location, or conditions in an environment. Changes can occur to population counts, identities, thematic attributes, spatial, or temporal characteristics.
Movement	A change of place or position; a progress, change, development; passage from place to place, or from one situation to another.	Shift in location of a geographic object over time. The object must maintain the same identity during a movement.

In the ever-changing world, geographic dynamics is ubiquitous, and the concept of geographic dynamics is fundamental to scientific inquiries and to decision making at all levels. Our daily schedules are filled with spatiotemporal activities, such as trips that require planning under the constraints of space, time, resources, and opportunities. Individually, spatiotemporal reasoning helps us plan itineraries to accomplish these activities. On the other hand, for a group of individuals, meeting arrangements need coordination of collective spatiotemporal activities to formulate solutions that are feasible. Generally speaking, activities are commonly purpose-driven.

The goal-seeking nature of activities is likely to reflect on spatiotemporal patterns of activities as resources and opportunities necessary to reach the goals. The Severe Acute Respiratory Syndrome (SARS) global outbreak in late 2002 and early 2003 is an example that demonstrates the importance of understanding dynamics in tracking spatiotemporal pathways of individual activities to determine potentially infected people and identify the original hosts. Similarly, much can be learned about drug traffickers and extremists through spatiotemporal analysis and modeling of their activities in order to understand relationships among these activities and the environment that provides resources and opportunities to nurture such activities. Moreover, we can also glean ways to coordinate and promote activities in order to reach a desirable outcome. A recent study of ants, for example, attempts to delve emergent patterns for traffic management and crowd control (Dussutour et al. 2004). Spatiotemporal modeling techniques, especially cellular automata, agent-based modeling, and genetic algorithms, have been broadly applied to solve the nonlinear problems that address emerging collective patterns from individual behavior and activities.

Events refer to happenings in space and time, such as traffic accidents, car bombs, or thunderstorms. In contrast to objects, events are occurrents; they happen and are then gone (Grenon and Smith 2004). The Geospatial Event Model (GEM) offers a formal model to reason about objects, events, and their settings (extents in space and time) in geographic domains, giving events the same level of importance as objects (Worboys and Hornsby 2004). The ephemeral nature of events signifies temporal importance; in fact, events typically demark times of significance and are often used as temporal references, even in information systems, where events mark changes in states. Events like sunrise and sunset, payday, and World War II commonly serve as relative time frames at different temporal scales.

A temporal frame of reference is also suggested and implemented in the Event-Based SpatioTemporal Data Model (ESTDM) by Peuquet and Duan (1995). Identification of significant events and their temporal relationships are often the first step in formulating hypotheses to explain eventual correlates and potential driving forces. Climatologists seek spatial and temporal correlations among climate events to explain dynamics at a larger scale, such as global climate change. Archaeologists examine artifacts to determine events that signify changes in life practices in space and time in order to retrospectively understand human cultural revolution. Furthermore, reduction of threats to our homeland security depends very much upon the understanding of how precursors and related events occur in space and time. The spatiotemporal patterns of unusual events (e.g., outliers) signal states of caution and are critical to hypothesis building to find the root causes and to probe means for mitigation and emergency preparedness (Cole and Hornsby 2005).

        While events refer to happenings, processes emphasize a "becoming" (e.g., initiation, transition, development, and evolution). A rainfall in a city, for example, is an event, but how a rainstorm develops and produces rain over space and time is a process. On one hand, a process may be considered as spatiotemporal profiling of an event by examining why the event occurs, how it unfolds, and what results from it. Consequently, analysis is directed to identify initial conditions, transitional drivers, developmental phases, and the life cycle that collectively characterize the process within the happening of the event. On the other hand, an event can be considered as an external force to a process that responds with changes through a series of states, phases, or steps. For instance, a river may change from erosion to deposition dominance after a landslide event happening upstream, and the introduction of a new immigration bill may alter the process used by human smugglers when transporting illegal immigrants into the United States.

II.   DRIVERS OF GEOGRAPHIC DYNAMICS

Geographic dynamics is complex, and its drivers (i.e., activities, events, and processes) may not be clearly distinguished. Activities of purchasing can also be considered as events that exchange money and goods. Likewise, a traffic accident event can be viewed as a process in which two vehicles move closer and closer and eventually collide with each other. For the purpose of this book, the distinction between activities, events, and processes lies in the focus of an observer. For activities, the observer tracks what individuals do in space and time, such as a person going to a train station or stopping one's vehicle to get gas. Activities are goal-oriented. Consequently, contextualizing spatiotemporal patterns of activities in geographic conditions can suggest the purposes of individuals. Routine activities often display some cyclic patterns (Campos and Hornsby 2004). Frequencies and patterns construed by routine activities form the baseline or normal condition. When an event happens, the normal condition is disturbed. As individuals take actions to cope with the disturbances, the spatiotemporal patterns of their activities change accordingly. Depending on the level of disturbances, these changes can be temporary or permanent.

Hägerstrand's Time Geography developed in the 1970s has been rejuvenated through the growing research interest in human activity patterns (Kwan 2004). Mei-Po Kwan, a plenary speaker at the workshop, illustrated GIS applications for analyzing the space-time paths of women with different ethnicities and demographic characteristics. Hybrid activities and embedded meanings became apparent when spatially contextualizing these activity paths. For example, based on a study of women's movements in Portland, Oregon, Kwan identified that African-American women are much more geographically confined when compared with Asian and Caucasian women. Her research on the daily activities and feelings of Muslim women after the 9/11 event revealed "a geography of fear" through the generation of space-time trajectories based on surveys in Columbus, Ohio that indicated where Muslim women feel most threatened. Furthermore, Kwan employed spatial interpolation techniques to estimate a density surface of activity opportunities among men and women and among women with part-time and full-time occupations. Such research provides insights into how different populations utilize geographic space and their flexibility and constraints in planning and taking spatial activities. Figure 2.1 and Figure 2.2 show examples of Kwan's research on space–time paths of human activities.

With respect to events, the observer looks for happenings in space and time. The happenings result from at least three possibilities. First, spatiotemporal coincidence of individuals causes something to happen, such as a traffic accident stemming from vehicle collision or a group of people meeting at a coffee shop. Secondly, happenings can also result from synergistic interactions among conditional factors in space and time (Stewart Hornsby and Cole 2007). For example, flooding results from coordinated spatiotemporal interactions among rainfall, terrain, soils, and land cover types. A bombing demands that all essential elements function as planned in space and time.

The third possibility of happenings is through connections of individual activities to form something beyond what individuals do. In other words, an event can occur without co-location or synchronous presence. A teleconference can take place as long as participants are connected through a communication means that enables the meeting at distributed locations. Furthermore, distributed and asynchronous Internet connections enable individuals to join a Web discussion at different locations and different times. Spatial and temporal constraints in communication and how the constraints become relaxed by technologies are well analyzed by transportation geographers (Janelle 1995; Harvey and Macnab 2000; Miller 2005) as summarized in Table 2.2.

Table 2.2: Spatial and Temporal Constraints to Communication (Miller 2005, Based on Janelle 1995 and Harvey and Macnab 2000)

Temporal	Spatial	
	Physical Presence	Telepresence
Synchronous	Synchronous Presence (SP)	Synchronous Telepresence (ST)
	Face to face	Telephone
		Instance messaging
		Television
		Radio
		Teleconferencing
Asynchronous	Asynchronous Presence (AP)	Asynchronous Telepresence (AT)
	Refrigerator notes	Mail
	Hospital charts	E-mail
		Fax machines
		Printed media
		Web pages

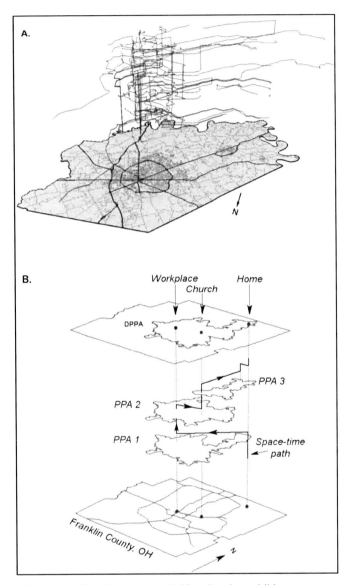

Figure 2.1: Visualize human activities. Panel A exhibits space–
time paths of activities from a given population. Panel B
illustrates space–time paths annotated with centers of activities
and their linkages (Kwan 2004).

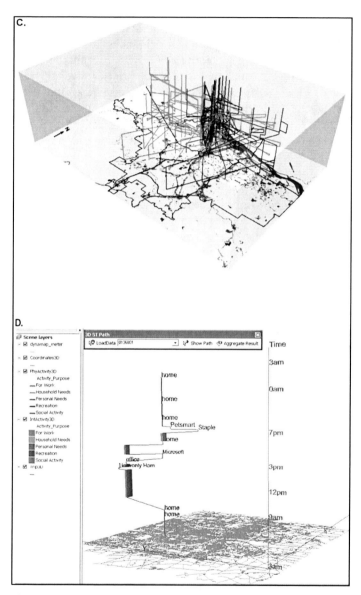

Figure 2.2: Panel C shows an aggregation of space-time paths from multiple populations (Kwan 2004). Panel D presents the software developed by Kwan's group for the visualization and analysis of human activities (Kwan workshop representation).

Beyond communications, there are environmental events afforded by connections among distributed and asynchronous activities. An El Ninõ event, for example, connects the irregular cycle of shifts in ocean and atmospheric conditions to abnormal weather that causes severe droughts and wildfires in Indonesia, Australia, or southeastern Africa, and floods and mudslides triggered by intensive rainstorms in Ecuador, Peru, and California. A terrorist attack, nevertheless, requires all types of activities (SP, AP, ST, and AT in Table 2.2) with careful coordination among training, materials, and actions.

With respect to *processes*, the observer focuses on the progress or transition and usually attempts to identify distinctive phases in development. Apart from activities and events, processes proceed continuously and gradually in space and time. Landscape change is one example of a process. Changes to a landscape can be regarded as progressive even though certain changes can be abrupt, such as a volcanic eruption; how the process of the eruption changes the landscape is progressive (Figure 2.3).

While a catastrophic event can result in instant impacts on the environment and society, a long-term process can produce large-scale, significant, lasting consequences at an unperceivable pace. Climate change is a good example of a seemingly slow process with broad and considerable threats to the human environment. Likewise, cultural, and social processes that nurture extremist ideologies may progress slowly, but ideologies can promote many to commit terrorist acts. Defeating an ideology may be much more difficult than defeating a physical threat. A process may be stimulated or modified by a series of activities or events during its development. A process can also trigger events and promote activities.

In his plenary presentation, Michael Goodchild, conceptualized processes as rules and equations that are nomothetical and universal. He proposed the use of GIS databases to capture geographic forms and the use of software to model processes. Indeed, processes are commonly represented by system diagrams (Figure 2.4) or functional relationships (Figure 2.5) to capture interactions among components under certain conditions. These functional relationships and mathematical models are theoretically, experimentally, empirically, or computationally determined.

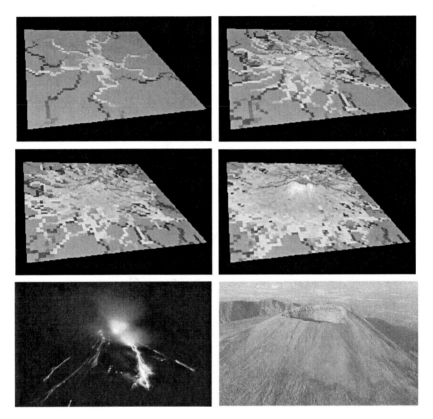

Figure 2.3: (See color insert.) Modeling of volcanic eruption (Raymond Sluiter; Maguire et al. 2005).

Both physical and human geographers as well as other researchers often relate forms, patterns, and processes to address geographic dynamics in the classic process-based approach to unite space and time. Specifically, "every empirical concept of space must be reducible by a chain of definitions to a concept of process" (Blaut 1961). The legendary geomorphologist, William Morris Davis, developed the Davisian model of fluvial landscape evolution that connects forms to the fluvial processes and characterizes fluvial landscape dynamics into three cyclic stages of youth, mature, and old age (Davis 1889). Sauer's school of cultural-historical geography stressed the importance of research to record, illustrate, and synthesize geographic patterns and explore geographic processes (Sauer 1956). In doing so, historical and cultural geographers build a "retrospective science" of cultural and historical processes in geography that enables us to acquire an ability to look forward (Sauer 1941). Geographic dynamics pervade over space and time, and this pervasiveness empowers the past as the key to understanding the present and predicting the future.

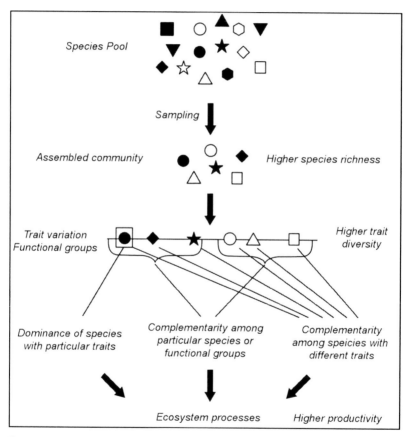

Figure 2.4: An example of using a system diagram to show the mechanisms involved in processes that select species and change biodiversity in an ecosystem (Loreau et al. 2001). Reprinted with permission from AAAS.

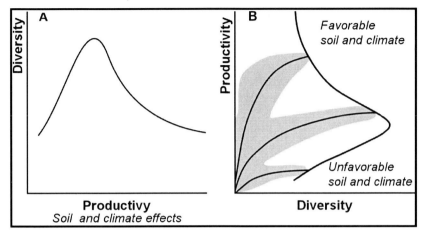

Figure 2.5: An example of using functional relationships to indicate the nature of involving processes. Panel A shows diversity as a unimodal function of productivity under changes in environmental conditions at a regional scale. Panel B shows the effects of biodiversity on productivity at a local scale under three environmental conditions as shown by the three shaded curves in B. The vertical curve is the same curve in Panel A (Loreau et al. 2001). Reprinted with permission from AAAS.

III. OUTCOMES OF GEOGRAPHIC DYNAMICS

Changeand movement are observables of geographic dynamics. Without change or movement, there is no means to detect or measure geographic dynamics. Perception of change implies two assumptions. First, the observed remains the same identity from T1 to T2. Identities allow the observer to recognize and trace the same objects for change analysis. Hornsby and Egenhofer (2000) proposed an identity-based change model with a set of primitives and operations. The set of primitives include identity states of objects and transitions. There are two basic identity states of objects: existence and nonexistence. Transitions link identity states of objects to express the progression of an object from one state to another. Moreover, the model attaches two qualifiers: with history and without history, to existence and nonexistence. Operations then can be used to determine the types of changes based on one of the four possible identity states to another.

A.   Change: Issues on Identify and Change

For certain continuants, like counties or buildings, determination of identities may be more certain than for other continuants or occurrents in geographic domains. Examples include rivers, traffic flows, money trails, and storms. When stream A and stream B merge, will it result in a new stream C downstream or should it be considered as one stream taking the other downstream? When a storm spins off two storm cells, should both the spin-off cells be given new identities, even if one cell appears dominant as a continuant from the original storm?

Conventionally, location is frequently used as the basis for identity in GIS databases. Both vector-based and raster-based models of space center on locations (Yuan 2001a). Vector GIS data models first determine an object's location by coordinates of points, lines, polygons, or other higher dimensional geometries and then determine the attributes at the defined locations. Object identifiers and geographic coordinates are required attributes in a GIS vector data set and must be determined before other attribute data can be assigned. Raster data models explicitly specify every possible location (cell) at the resolution of interest (cell size). The location-centric conceptualization well serves change-based reasoning in geography. Since location determines identity, comparison of attributes at a location over time reveals changes or no changes to the location.

Attribute comparison for change identification constitutes the second assumption that the observation centers on attributes by assessing if an attribute set is different from T_1 to T_2. Such differences may be interpreted as change in attribute values or change in attribute types. In addition to defining the significant level of change, the inherited spatial component in geographic domains brings additional complication to potential dimensions of change, including changes to geometry, spatial extent, topology, and internal structure.

Changes in geometry (or shape) likely induce changes in spatial extent (or size), but the reverse may be false. A geographic object can maintain its shape while experiencing expansion or contraction as long as the change is isotropic. Changes to the internal structure of geographic objects are only recently identified and addressed in GIScience literature (Yuan 2001b; McIntosh and Yuan 2005a). Some geographic objects possess spatial heterogeneity. For example, storms move as individuals, but a storm's properties, such as precipitation intensity and temperature, vary spatially within the storm (Figure 2.6). While a crowd may behave as a unity during a political demonstration, the crowd may be composed of several organizations with individual attempts and motives. Both the internal structure of geographic objects and changes to the internal structure relate to dynamics operating in and carried by the geographic objects.

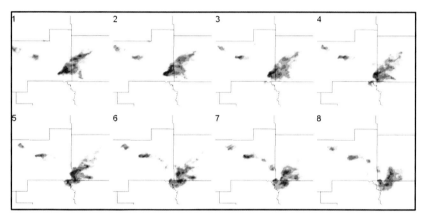

Figure 2.6: Examples of geographic objects with spatially variant properties. Precipitation intensity varies spatially within each storm cells. The panels are National Weather Services NexRAD images in a time sequence of severe thunderstorms over Greer, Washita, Beckham, and Kiowa Counties in southwest Oklahoma (images assembled by Nicholas Benson).

B.   Change: The Case of Networks

Changes to an object's internal structure can also result from spatial variations of its properties and to its topology. Topological changes alter the connections among elements of an object. Steven Prager, a participant in the workshop, introduced the use of networks to model political relationships among nations. Changes in the topological structure reflect the evolving nature of the network (Huntington 1996). In addition, changes in flight networks in the United States were also examined to demonstrate how the changes in these networks influence flight density (Figure 2.7).

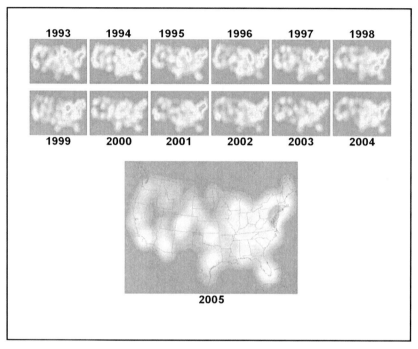

Figure 2.7: (See color insert.) Change in flight density as a result of topological changes in flight networks (presented by Steven Prager at the workshop

Analysis of changes in network topology applies to social science research with an emergent interest in developing and using Network Theory to interpret the changing structure of scientific practices and social relations. The creative enterprise in science has been evolved from distinguished individuals, such as Charles Darwin and Albert Einstein, to collaborative teams (Figure 2.8). A change in network structures results in changes in the roles and functions of individuals in a scientific enterprise as well as the outcomes, which evolve from predominantly single-authored papers to team publications (Barabási 2005).

In both physical and social worlds, network topology indicates functional dependency among elements and the overall purposes of the network. The topology of a hydrological network, for example, is influenced by the topographical and geological structures where the streams flow, and the network topology determines drainage efficiency in a watershed. Changing the topological structure of a network alters associations, functions, and operations in the network. Network characteristics change through evolution driven by internal and external forces.

In general, two principles may guide the evolution of a drainage network. First, the statistical principle (or the Principle of Randomness) is derived by applications of random walk techniques that result in a drainage evolution into random, highly probable organization of drainage networks (Abrahams 1984). The idea of high probable organization stresses the minimum energy expenditure to the system and maximum energy expenditure within the system (Shreve 1975). Second, the optimization principle is also known as the Principle of Least Work, which suggests that the development of a drainage network is not a random process but is determined by physical laws of conservation of energy and energy distribution in an open system (Rodriguez-Iturbe et al. 1992). In a drainage system, the physical laws of energy conservation regulate the balance between the energy consumption per unit area of overland flow and the energy consumption per unit length of channel flow. Circular drainage in forms of a honeycomb-like hexagon has been shown to offer the advantages of least work (Woldernberg 1969). Consequently, Woldernburg suggested that when first order streams exhibit hexagonal forms, higher-order drainage networks will also be least-work assemblages of nested hexagons. Debates over the randomness and deterministic approaches have enriched the research in drainage network evolution (Smart 1979).

Figure 2.8: The evolution of science enterprise from high-profile scientists and paired collaborators to multidisciplinary research teams (Barabási 2005).

Study of drainage network evolution provides three perspectives on dynamics to aid identification and interpretation of network changes, which may be generalizable to other types of networks: (1) guiding principles, (2) structural correspondence to energy balances, and (3) hierarchies. While distinctive sets of principles may be more appropriate for different types of networks, the statistical principle of randomness and the optimization principle of determinism may be broadly applicable to many network systems, solely or alternatively throughout the development of a network.

Some networks may be solely deterministic and others may be solely random. Some networks may form through random causes, but become more deterministic as the network grows, and others may develop from the opposite direction. The development of chat groups in cyberspace, for example, may start with random connections or interest among individuals, but become attractive to certain people when discussions become focused on certain topics or ideologies. On the other hand, a political demonstration may start with a deterministic structure in an orderly manner, turning random when unexpected events disrupt the order. Elicitation of guiding principles to network development provides clues to the central function of the network (e.g., communication or dissemination) and to the evaluation of network stability.

Structural correspondence with energy balance and hierarchical structures are the result of the respective guiding principle of networks. This premise assumes that a network will adjust its structure over the course of its development to meet the needs of energy balance. Structural adjustments can be topological or hierarchical. A transportation network, for example, often includes hierarchical ordering of highways and streets to connect locals to regions. Depending upon the distributions of locals and political or commercial needs, transportation networks can be radial, gridded, hubs and spokes, or other forms of connectivity. Both topologically and hierarchically, the optimal structure of hexagons in drainage networks interestingly matches the six degrees of separation in social networks (Milgram 1967). Is it a coincidence or is there an implication of six in network structures? Are triads the fundamental unit of a social network (Scott 2000)? Could there be a connection between the unit of three to the structure of six in network development?

Social networks associate people by relationships and communication in various structures. As shown in Figure 2.9, terror networks connect individuals to organizations generally in three basic topologies: (1)

the chain network in which a line connects people in a sequence, and end-to-end communication must rely upon intermediate connections (this type of chain network is common in smuggling); (2) the star, hub, or wheel network requires connections through the center and is popular for tactical operations; and (3) the all-channel network that connects everyone to everyone else promotes collaboration and is popular among militant small groups.

Hybrid networks that include two or all the three topological structures are common, and many networks are dynamically reconfigured throughout operations to perform different functions throughout the development of the network and the process to complete a mission. Many interesting research questions arise to understand network dynamics. Are the guiding principles in hydrological network development applicable to explaining and predicting dynamic reconfiguration of social networks? Are there principles unique to social network development? Can universal or specific laws of network development be formulated to identify stages and thresholds of network development and predict stability and sustainability of a network?

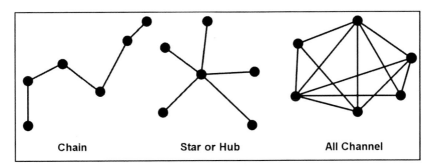

Chain   **Star or Hub**   **All Channel**

Figure 2.9: Three basic network topologies.

C.   Movement

The above discussion elaborates on how geographic dynamics manifest themselves through changes to attributes or structures (such as networks) over time. Through identifying attribute and structural differences, we gain an understanding of the developmental discourse of geographic objects, phenomena, and networks. Additionally, movement constitutes another observable of geographic dynamics. Movement is observed by shifts in location with or without attribute or structural differences. Similar to the perception of change, recognition of movement rests on the premise of the abilities that geographic objects can maintain individual identities and the observer can discern objects' identities. As mentioned earlier, spatial locations provide a convenient basis for geographic identities. Spatial locations can be modeled as point locations, line locations, or area locations that include spatial extents. Many change analysis tasks take advantage of location persistence and define geographic objects starting with spatial locations of points, lines, or areal representations. Compared to change, movement introduces a new dimension of complexity because geographic semantics, not spatial location, should serve as the basis for analysis.

Geographic semantics is defined here as the core meaning that is associated with a geographic object, i.e., the combined essential properties and elements that constitute a geographic object and that can be used to identify the geographic object. It is not uncommon for geographic semantics to vary depending upon perspectives and domain interests. Forests, savannah, and wetlands all have multiple definitions. The use of the term geographic semantics is to retain the flexibility to interpret the appropriate core meaning in identifying a geographic object (Yuan 1996). Once a geographic object has been identified, we can track its movement at the same granularity or track it over multiple granularities (Hornsby and Egenhofer 2002). With the popularity of global positioning systems (GPS) technology, many movement data are collected as a time-series of x-y coordinate pairs. GPS-based tracking has become popular in analysis of human activities and public health (Kwan 1999; Kwan and Schuurman 2004). In addition to GPS technology, research on animal home range and dynamics applies tagging techniques to track movements (Mass and Yuan 1999).

For geographic objects with geometries of one or more dimensions, tracking movement can be challenging. Movement of linear, areal, or other higher-dimensional objects cannot be fully analyzed through a set of points over time. These objects may possess spatially varied properties as fields. Geostatistical methods have been an important tool in modeling spatiotemporal fields (Holawe and Dutter 1999). Storms, wildfires, and pollutant plumes are examples of such field-objects (Yuan 2001b). Hence, no single points can be used as representatives of these field-objects (see Figure 2.6). Furthermore, not all parts of an areal object move at the same speed and speeds at some part of the object may change over time. Determination of the path and speed of a movement can again be challenging as there is no movement at a single point representative of the entire object. The speed difference forms sheer force that creates rotation or vorticity inside the object, such as a hurricane. When the sheer force within an object exceeds the ability that the object can extend itself, it is likely to split into two objects. Meanwhile, two objects may move, collide, and merge again into one. Another potential outcome is that one of the two objects crashes into multiple objects and both objects become disseminated. Therefore, tracking these complex geographic objects requires a suite of indices that measure individuality of these objects, internal structures and behaviors of embedded fields, group behaviors of object aggregates, and spatiotemporal transition of objects (McIntosh and Yuan 2005b).

In addition to GPS survey points and remote sensing imagery, video data have become another primary source for a wide range of movement analysis from tumor growth and muscle movement to animal behavior. Shape and kinematics are important clues for object recognition, matching, and registration from videos (Veeraraghavan et al. 2005). The shape-based approach relies on identifying the shape of an object and the deformation in a sequence of the shape to extract motion. The kinematics-based approach applies dynamic affine invariant or potential trajectories to determine an object moving across video frames. Both approaches appear rather robust in tracking movement across various applications, including surveillance in human behavior analysis (Ou et al. 2005), fish movement (Lucas and Baras 2000), insect tracking (Noldus et al. 2002). However, incorporating content-based analysis and latent semantics approaches is important to resolve ambiguity and retain association for object recognition and tracking (Hsieh and Huang 2002; Sivic et al. 2006). The importance of video analysis of movement (as well as change) grows as increasing applications are seen in medicine, surveillance, behavior, journalism, and many more.

IV.   SUMMARY

This chapter examined dynamics in geographic domains from a system's perspective. Activities, processes, and events are drivers that introduce energy and materials into a geographic system and induce system adjustments through change and movement. These three drivers are scale dependent, and their differentiation may be situationally determined. Rainfall processes become rainfall events when observations shift from following the evolution of a rainstorm to emphasizing a particular experience of a location. An event or a process can be considered as an activity when the focus is placed upon actors who perform the activity. Likewise, a process may consist of many activities (such as the process of social network development and intermediate activities to create contacts and build relationships). An event may consist of many processes such that a commencement includes processes of processions, speeches, degree presentations, and other routines.

Complexity of geographic dynamic arises from the potential interwoven relationships among the three drivers and their scalar dependency resulting hierarchies. Accordingly to Hierarchy Theory (Warren 2005), higher level drivers in a hierarchy operate at a lower frequency and with a greater spatial and temporal influence to the system of geographic dynamics. Higher level drivers also impose boundary conditions to drivers at lower levels. Meanwhile, lower level drivers operate at a higher frequency and supply energy and materials necessary to sustain higher level drivers. Since the Hierarchy Theory considers that spatiotemporal scale is the factor to level identification, the hierarchical relationships among drivers demand that the understanding of dynamics at one scale requires analysis of drivers at three scales: the focal level, one level higher (the super level), and one level lower (the sublevel; Ahl and Allen 1996). Single scale analysis of geographic dynamics overlooks the lower and upper contexts that afford and constrain the dynamics for sustainability.

Geographic dynamics is observed through change and movement. Change addresses differences in properties or structures. Movement can be regarded as change in locations, but movement also prescribes speed and direction. Determination of identity is crucial for analysis of change or movement. GIS technology commonly uses location (be it a point, a line, or a polygon) as the basis for identity. While location-based identities facilitate change analysis through spatial overlays, the approach inherently has identities fixed with location and loses capabilities to represent movement. Object-based identities are challenging as geographic objects are often not confined to a constant shape nor do they remain unities over time. Many geographic objects can split into two objects or merge into one object. Many of them have multiple definitions depending upon perspectives. New developments in video analysis and tracking algorithms can probe new GIS methods to identify movement in time-series data. However, effective handling of splitting and merging geographic objects remains a research challenge when understanding geographic dynamics.

GIS researchers have been diligently examining and proposing new ways to handle change and movement since the late 1980s (Langran and Chrisman 1988), comparable to time research in database communities (Allen 1983; Snodgrass and Ahn 1986). Researchers from both temporal GIS and temporal database seem to follow a parallel development that applies *time-stamps* to tables, tuples, and values (Yuan 1999). Around the mid-90s, GIScientists have recognized the importance of activities, events, and processes, and many advocate the GIS data models based on these dynamics drivers. The next chapter aims to provide perspectives from data modeling and analysis to elaborate on GIS handling of geographic dynamics.

CHAPTER 3

DATA MODELING APPROACHES FOR GEOGRAPHIC DYNAMICS UNDERSTANDING

- *How can geographic dynamics be represented in Geographic Information Systems?*

- *How are these representations used to support reasoning, querying and analysis of geographic dynamics?*

Time and space are essential dimensions to geographic dynamics. The multitude of complexities arising from geographic dynamics demands robust GIS representation schemes that can account for the three drivers (i.e., activities, processes, and events) and two observables (i.e., change and movement) as discussed in the previous chapter. Representation of space in GIS typically follows the cartographic tradition, particularly in forms of choroplethic maps. Cartographers use maps to represent geography in two-dimensional media, pieces of paper or a wall, for example. Adopted in GIS, such a static two-dimensional spatial framework (commonly known as data layers) poses many challenges to the development of a temporal GIS capable of handling geographic dynamics. The similar time-stamped approaches adopted in temporal database management systems (DBMS) and temporal GIS suggest the complex nature of dynamics (with or without involvement of space) and the difficulty in capturing temporality in digital representations (Figure 3.1).

I.   TIME-STAMPED APPROACHES

GIS data layers naturally promote the use of time-stamped snapshots to store spatial data over time. At every time of interest, a complete new data layer is created to capture the state of geography in an area. When ordered by time of

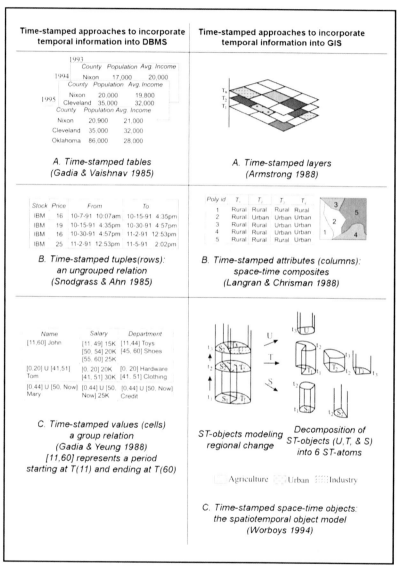

Figure 3.1: Time-stamped approaches to incorporate temporal information into DBMS and GIS databases (assembled figures from Yuan 1999).

acquisition to form a temporal sequence of snapshot data layers, a GIS captures the area's history; i.e., how properties of a given area change over time. An obvious drawback is data redundancy, especially in a relatively stable area where changes rarely occur. Another shortcoming relates to the system's inability to recognize dynamics and to miss changes that occur between snapshots, making this approach less useful for supporting information queries about change and movement. While snapshots are considerably limited in supporting spatiotemporal query and analysis, the approach remains popular in most GIS applications because of its simplicity and compatibility with many spatiotemporal data acquisition methods, such as satellite remote sensing, aerial photogrammetry, and ground-based environmental sensing and observations.

To eliminate redundancy due to repeated data of constant values in snapshots, the space-time composite model condenses spatial and temporal data into a set of maximum homogeneous units in space and time (Langran and Chrisman 1988). Every space-time composite unit, therefore, represents an area of uniform change during the period of interest. The model's major drawback relates to data updates that require reconfiguring space to account for changes occurring within a previously identified space-time composite unit. As the space-time composite unit is no longer homogeneous, it is split into two space-time composite units to ensure homogeneity of history within each unit.

When space is reconfigured, data identifiers are reassigned to the newly formed space-time composite units. There is no mechanism, however, for maintaining knowledge about how the previous space-time composite units relate to the new units. Another early effort at spatiotemporal modeling is the Space-Time Object Model. In this model, time is treated as an independent dimension orthogonal to space, and linkages among spatiotemporal homogeneous units (space-time atoms) are built to recognize a space-time object useful for showing how the spatial extent for a given spatial attribute type (i.e., a constant attribute value, such as agriculture land use) changes shape and size over time (Worboys 1994). Time-stamps are applied to denote the beginning and the ending of a given spatial extent.

II.   CHANGE-BASED VERSUS EVENT-BASED APPROACHES

Temporal DBMSs or GISs designed around the concepts of time-stamps often involve cataloging temporal data to encode historical records for archival purposes, e.g., for storing data about cadastres or zoning records. From these efforts, change becomes a major focus of analysis, although for many applications, merely recording change at locations is insufficient and ineffective, because information about the patterns of change through time is essential. To this end, time can serve a more effective reference than space to manage data about change.

The Event-Based SpatioTemporal Data Model (ESTDM) organizes locations of change in a time series to facilitate temporal data retrieval and analysis (Peuquet and Duan 1995). In this model, "events" are used to reference times of change within a raster data layer. A base map represents the initial condition. A header file records a temporal sequence of events when changes occur, and components encode the locations where changes occur as well as the attribute values at the event time. For example in Figure 3.2, a change occurs to the shaded area with a new value of 15 corresponding to the event time t_i. The area of change is encoded in a component with a location list of (x, y_1, y_2) to indicate the location of cells where the change occurs in the raster data matrix (Figure 3.2). From the event list, the ESTDM can readily provide information where changes occur at a given time and facilitate comparisons of changes and change patterns at different times.

More recently, events have been used to index time-dependent data for adaptive visualization (Figure 3.3). One of the workshop plenary speakers, William Ribarsky, presented visual analysis of urban landscape dynamics with an emphasis on urban legibility that makes a city comprehensible and navigable. His visual analysis approach aims at understanding what should be rendered and at what level of details in order to effectively support navigation in an urban environment. Figure 3.4 is a simulation showing what an individual would see when the individual navigates through a complex urban environment. In the simulation, Ribarsky demonstrated the need to depict the city around the concepts of paths, districts, nodes, and landmarks. He applied a time event structure to embody these concepts that adaptively ensure the city is legible and recognizable at all scales by applying a global hierarchical data structure that organizes data at multiple spatial resolutions and from multiple data sources. The time event structure effectively guides the global hierarchical data structure to dynamically display urban landscape in terms of paths, districts, nodes, and landmarks at a scale appropriate to what an individual can see when navigating through the city.

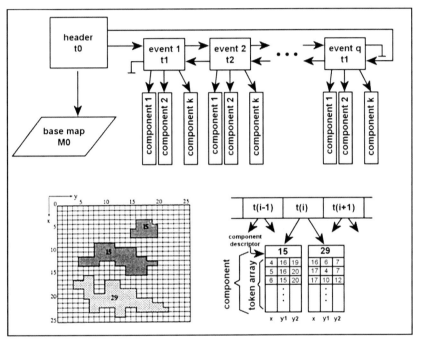

Figure 3.2: The data structure of the event-based spatiotemporal data model (Peuquet and Duan 1995).

The need to represent events in data models is one of the recent research foci in geographic information science. Worboys (2005) discusses the development of temporal snapshots, object lifelines, and event chronicles and argues for an equal treatment of things and happenings in GIS databases. Ontologically, things (e.g., buildings or roads) can be characterized as continuants that endure through time, whereas happenings are considered as occurrents (e.g., storms or protest marches) cease to exist after a certain period of time (Grenon and Smith 2004). In data modeling, an important consideration is to distinguish between what exists in reality versus how this reality is represented in a GIS. While many technical terms remain loosely defined in GIScience, Worboys and Hornsby (2004) made the following distinctions: An entity refers to something in reality, and an object is a data modeling construct representing an entity in reality. Likewise, an occurrent refers to a happening in reality, and an event is a data modeling construct representing an occurrent in reality. To avoid confusion, we will specify data objects as representation of entities and data events as representation of occurrents.

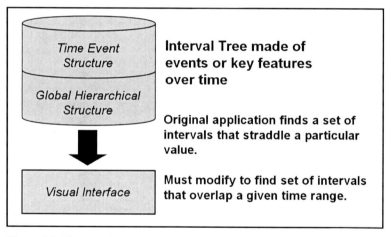

Figure 3.3: Time event structure to support 3D adaptive terrain visualization (from Ribarsky's presentation at the workshop).

Object-orientation is a popular approach used to model geographic entities in reality and frequently adopted in temporal GIS design (Wachowicz 1999). Advocating for event-oriented approach emphasizes the need to represent real world occurrences as data event objects and support information query about event attributes and event relationships. With this support, a temporal GIS will be able to facilitate the treatment of underlying events rather than temporal references to objects.

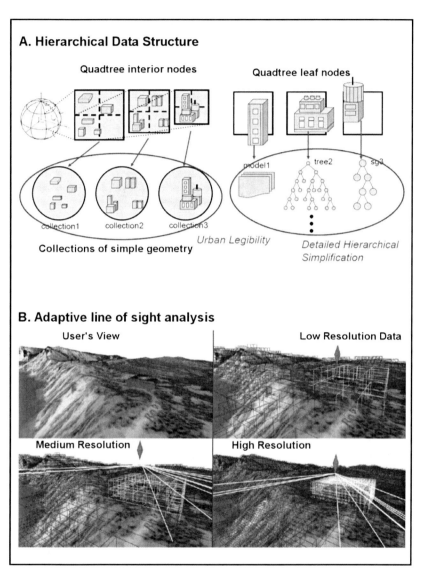

Figure 3.4 (See color insert.): An adaptive multi-resolution line of sight visual analysis of an urban environment (represented by William Ribarsky in the workshop).

In summary, time-stamp approaches center on the concept of change to incorporate temporal components into spatial data. Change can occur to geographic features or be perceived by a moving observer. When change occurs, a time-stamp denotes the time of change. In essence, a time-stamp is a temporal index to spatial data. Event-oriented approaches, however, focus on the dynamic happening as a whole, and not just the time of the event, allowing event attributes and relations to be distinguished in addition to the mode usual in object attributes and relations.

III.   MOVEMENT-BASED VERSUS ACTIVITY-BASED APPROACHES

Besides change, movement is the other measurable of geographic dynamics. Some temporal GIS applications, such as navigation and shipments, need to be movement centric. One can argue that movement is just change in locations. The concept of change can be applicable to movement and, therefore, change can be considered as the only observable of geographic dynamics. However, location is exogenous to any mobile entity and, therefore, should be considered separately from property changes.

Movement involves two essential elements: an individual with a persistent identity and a shift in locations over time. For entities that can be represented by point-based data objects, determination of a location shift is much more straightforward than those represented by linear or areal data objects. Most of movement analysis studies are based on point-based data objects. In the workshop, Darren Scott presented a project on predicting activities and travel behavior in urban areas (Figure 3.5) in which individuals are represented by data objects modeled as points, and movements of individuals are represented by line sequences to denote moves from origins to destinations. Similarly, Colin Ware, one of the workshop's plenary speakers, presented a study of tracking the movement of a humpback whale. In addition to tracking the whale's movement, attributes along the movement are also recorded so that movement behavior can be discerned from analysis. In Figure 3.6, for example, the sawtooths indicate the occurrences of fluking, and the yellow segments indicate the whale rolled over at 40-degree angles from the horizon. The figure shows that the whale tends to roll at the bottom of its dives.

Global Positioning Systems (GPS) technology has promoted research in moving objects modeling and analysis in GIS. Applications include transportation, navigation, epidemiology, location-base services, and many more. In GIS database modeling, geospatial lifelines are used to represent "a continuous set of positions that are occupied by an object in geographic space over time" (Mark and Egenhofer 1998).

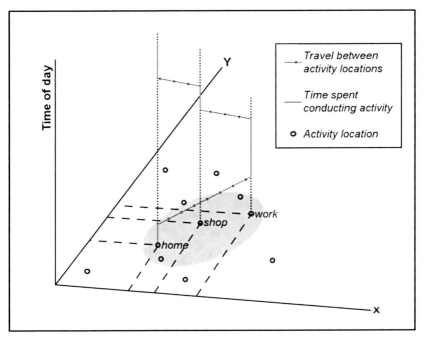

Figure 3.5: Apply space-time prism constraints to examine movements within a scheduling process (presented by Darren Scott in the workshop; Meyer and Miller 2001). Copyright The McGraw-Hill Companies, Inc.

Fundamental to the lifeline data model is the study of Time Geography (Hägerstrand 1970) in which the paths individuals are taking are the basic data constructs. Geospatial lifelines may be determined based on daily activities to evaluate how people individually or collectively use an environmental (or urban) space (Kwan 2004) or based on addresses to examine issues of potential impacts of environmental exposure (Mark and Egenhofer 1998). In either approach, movements and activities attributed to the movements of individuals constitute the data objects used to record the information in a GIS database.

Movement analysis relies upon the granularity of a geospatial lifeline (Hornsby and Egenhofer 2002) or distance between geospatial lifelines (Sinha and Mark 2005). When further analyzing the components of a geospatial lifeline, Hornsby and Egenhofer (2002) proposed lifeline threads to represent a linear approximation of an individual's continuous movement based on an ordered sequence of space-time samples and proposed lifeline beads to represent the possible space-time extent to which an individual can reach under space-time constraints. Comparing the lifeline beads of multiple individuals may suggest the potential interactions among individuals in meetings. Shifting between coarse and fine granules enables different views of movement patterns and spatiotemporal details of activity extents.

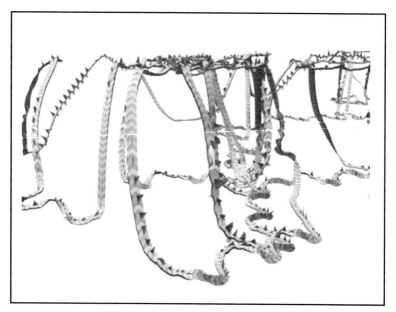

Figure 3.6: (See color insert.) The movement of a humpback whale (presented by Colin Ware in the workshop).

An alternative view of geospatial lifelines is considering multiple point objects with temporally ordered discrete points sampled at regular or irregular spacings. Based on attribute distributions along neighboring points along a lifeline, interpolation routines can be developed to fine tune a lifeline's temporal granularity or generalize the lifeline pattern at a coarse grain. Laube and colleagues explore the multiple point objects (MPO) idea to analyze relative motion within groups of movement objects (Figure 3.7). These basic types characterize different kinds of movement, such as "many objects moving with equal speed" utilizing data on motion azimuth, speed, and change of speed (Laube and Imfeld 2003; Laube et al. 2005).

IV.   PROCESS-BASED REPRESENTATION

GIS data modeling must deal with actual processes, not just the geometry of space-time (Chrisman 1998). As studies consider the fusion or integration of space and time, representation of space and time cannot overlook the need to represent processes. In geography, nothing is purely space or time; everything is process (Blaut 1961). In the early part of this chapter, we state that events denote happening while processes mark transformation of stages or phases in space and time, but we also recognize that disagreements exist among researchers regarding the distinction between processes and events (Galton and Worboys 2005; Worboys 2005).

Graduate transformation suggests dynamics that propagates (continuous) or phases (stepwise) in space and time. In analyzing possible dimensions that dynamics can manifest, Goodchild et al. (2007) proposed a conceptual cube to identify eight basic types of possible change and transformations of geospatial phenomena, in addition to change to attributes (Figure 3.8). Attribute changes can occur to any of the eight blocks for analysis of geographic dynamics. The block representing static geometry, static internal structure, and static movement (lower left corner in Figure 3.8 carries geographic dynamics that only involves changes to attributes. On the other extreme, the block representing elastic, evolving, and moving (upper right corner in Figure 3.8) encounters the highest degree of geographic dynamics with possible changes to attributes, shape, internal structure, and location. Changes to topology are considered as one kind of internal structure changes because topology is a property of the whole, not of individual parts.

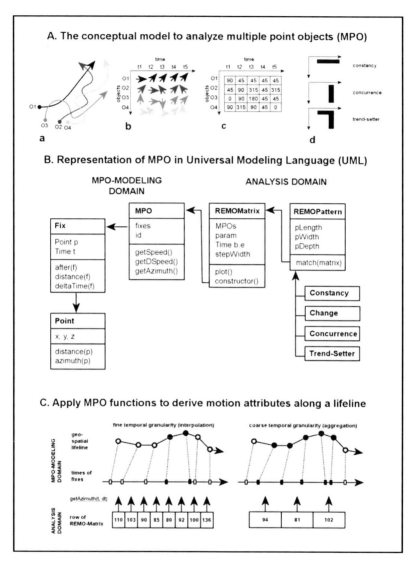

Figure 3.7: The multiple point object (MPO) approach to store and analyze information about movement in a GIS (Laube et al. 2005).

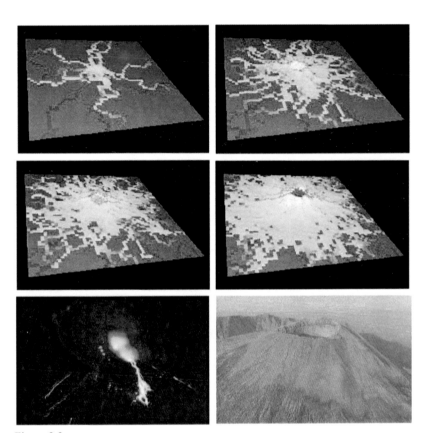

Figure 2.3

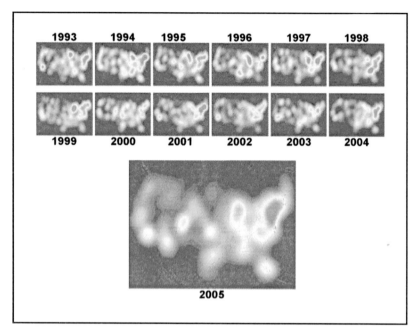

Figure 2.7

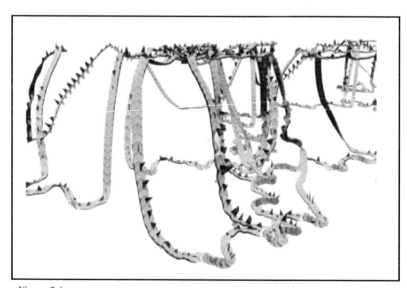

Figure 3.6

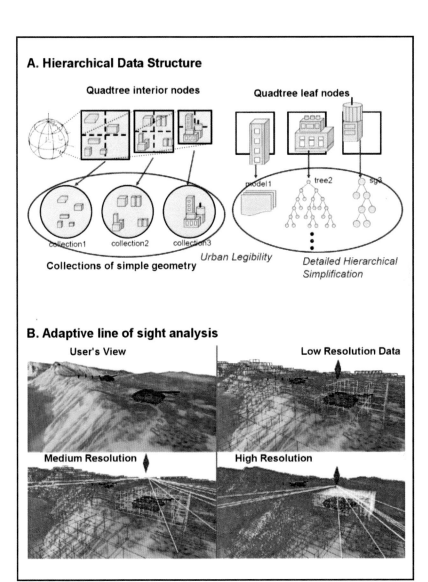

A. Hierarchical Data Structure

Quadtree interior nodes

Quadtree leaf nodes

model1   tree2   sg3

collection1   collection2   collection3

Urban Legibility

Collections of simple geometry

Detailed Hierarchical Simplification

B. Adaptive line of sight analysis

User's View

Low Resolution Data

Medium Resolution

High Resolution

Figure 3.4

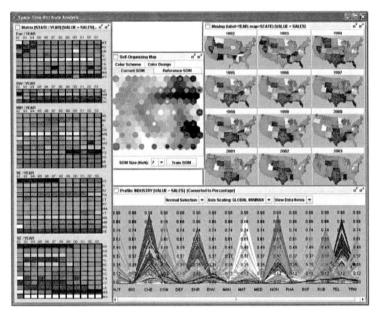

Figure 4.11

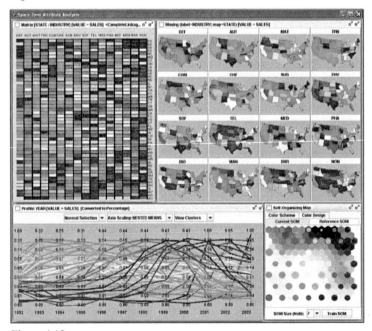

Figure 4.12

Comprehensive representation of processes will need to incorporate dynamic attribute, geometry, internal structure, and movement in GIS data models. The cube of geographic dynamics incorporates both observables (i.e., change and movement) and drivers (i.e., activities, events, and processes) into the properties of these eight data object types. Activity data objects emphasize the dimension of movement, but generally hold geometry and internal structure static. Event data objects represent occurrences that are attributed with uniform properties and no movement. In comparison, processes exhibit a high degree of complexity in geographic dynamics. Process data objects account for dynamics that involve in changes to attributes, geometry, internal structure, and movement during propagation and phases of process evolution.

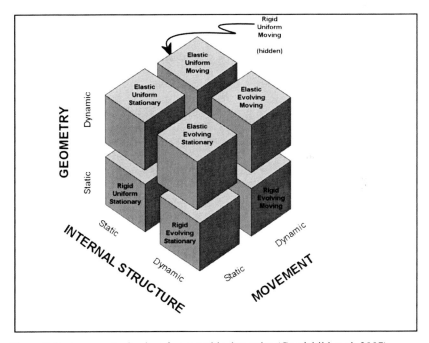

Figure 3.8: A conceptual cube of geographic dynamics (Goodchild et al. 2007)

As discussed in chapter 2, there are no universally accepted definitions of events and processes in the GIS community. Based on the definitions given in chapter 2, GIS representation of geographic processes should emphasize the transitions of phases from one state to another. The idea is used to represent storm processes through a hierarchical framework that captures footprints and transitions during a storm's development (Yuan 2001b; McIntosh and Yuan 2005a). The representation starts with zone data objects to represent spatial footprints of a storm at different times. The hierarchical framework links zone data objects over time to form a sequence data object that tracks the locations and geometries of the storm over space and time. To account for merging and splitting during the storm development, a process data object is used to embed joint sequences that result from splitting an original sequence or from merging two sequences. Finally, process data objects together form an event data project as long as the process data objects correspond to geographic processes that operate during a period of time, over a study area, and are justifiable by coherent dynamics, such as a weather system or a parade movement (Figure 3.9).

In the hierarchical framework, event data objects represent something happening and process data objects represent how something happened over space and time. As discussed in chapter 2, the distinction of events and processes depends on the spatiotemporal granules of interest. An investigation of why an event has happened requires the dissecting of the event into discrete or continuous processes to reveal the mechanisms and phases. An examination of a process requires identification of events denoting what have happened throughout the initiation, strengthening/weakening, phasing, or termination of the process. Therefore, an event data object may consist of multiple process data objects, and, likewise, a process data object may have multiple event objects in a GIS representation of events and processes.

Instead of direct representation of processes in data models, processes are generally represented by mathematical models or algorithmic procedures rather than data objects in GIS databases. Goodchild in the workshop also suggested the use of GIS data objects to represent forms and the use of equations to represent processes. Ultimately, decisions about whether or not to develop data constructs to represent processes depend on needs to access historical or modeled processes for analysis. Mathematical and computation models are indeed commonly used to represent mechanic processes or fluid dynamics. Statistical models are used to represent stochastic processes. More specifically, Random Walk, Markov Chains,

Cellular Automata, and Agent-Based Modeling are good examples of process representations. Random Walk models assume that moves taken at different locations and periods are based on random selections. Markov Chains expand a process into a sequence of state transitions. Cellular Automata and Agent-Based Modeling simulate processes of spread, reorganization, and movement based on decisions and behavior at individual levels. More discussions on mathematical, statistical, and computational methods to model processes will be discussed in the next chapter.

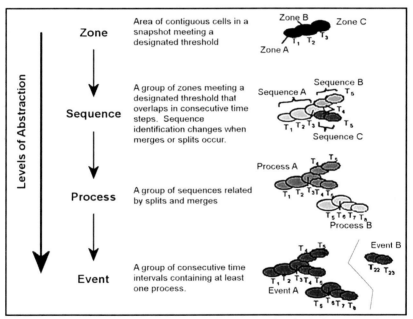

Figure 3.9: An example of process-based representation in GIS (McIntosh and Yuan 2005).

CHAPTER 4

VISUAL APPROACHES FOR GEOGRAPHIC DYNAMICS UNDERSTANDING

- *What are different ways to visualize geographic dynamics?*

- *How can different aspects of geographic dynamics be visualized and understood?*

Geographic dynamics is complex, multi-dimensional, and multi-scalar. Understanding geographic dynamics, consequently, demands comprehensive approaches to address histories, behaviors, consequences, relationships, interactions, and evolution in space and time. Methods have been and are being developed for understanding geographic dynamics. Some methods may be categorized as "storytelling" with emphases on replaying activities, events, and processes. Animation and exploratory visualization techniques fall in this category. Other methods center on "analytics" with aims to understand what, how, and why dynamics manifests itself in geographic domains. While both storytelling and analytics-based approaches are important to geographic dynamics understanding, a storytelling approach that animates images and time sequential GIS data layers in order to illustrate how geographic dynamics develops appears more popular because of simplicity and ease of creation and use. Meanwhile, many methods and tools for visual analytics have been developed to support exploratory data analysis (Gahegan et al. 2002), visualize data uncertainty (MacEachren et al. 2005), build a geo-collaboratory that communicates geo-scientific concepts (MacEachren et al. 2004; MacEachren et al. 2006) and many new exciting advances. In this chapter, we discuss both of these approaches for visualizing geographic dynamics.

I.   STORYTELLING VISUALIZATION

The storytelling approach refers to methods of replaying or forecasting visually the changes in dynamics of a geographic domain. This approach strongly relates to animation and can effectively draw insights into potential causes and interpretations of geographic dynamics. For example, Tobler's first law of geography (everything is related to everything else, but near things are more related than distant things) was inspired by animating a sequence of urban maps to show the decadal growth of the Detroit region from 1930 to1960 (Tobler 1970).

With the human's acute visual capabilities to recognize trends and patterns, animation frequently offers an effective means to recognize dynamic phenomena and perform mental projections into future development. There is no shortage of examples on the use of animation techniques that show, for example, global climate change or transitions in vegetation greenness indices using NASA imagery data (Harrower et al. 2000). Similar to the field-object approach in spatiotemporal data modeling (Yuan 2001) discussed in the previous chapter, additional insights of structural information about geographic dynamics can become more evident by integrating continuous raster fields and morphometric vector features than field- or object-based presentations alone (Rana and Dykes 2003).

Other examples of popular applications of animated visualizations include disease spread, migration, and hot spot shifts in crime occurrences. In some cases, the use of animated maps to visualize dynamic situations is the norm of communication. The Weather Channel, for example, commonly displays animations of temperature and precipitation maps to illustrate how a weather system is developing as well as predicted paths for the system. Increasingly, readily available geospatial data and Internet programming facilitates the creation of animated maps to communicate geographic dynamics at all levels. For example, the TimeMap Open Source Consortium develops time-enabling tools to create map animations that illustrate spatiotemporal distributions of Australian indigenous groups, the change of political geography over time in Southeast Asia, or the territorial changes of the Mongol Empire (1100-1400 CE), for example.[1]

---

[1] TimeMap and a suite of animated maps are available free at http://www.timemap.net

In addition to time dependency, animation-based visualizations can be created based on changing the observer's perspective. Fly-through visualizations are good examples. The landscape may be static, but flying over the landscape results in changing views of the landscape. The workshop plenary speaker, William Ribarsky, presented Visual Analysis of Urban Terrain in which navigation through a 3D urban terrain demands visualization capabilities to dynamically display urban landscape from different vintage points (see Figure 3.4 on page 37 and in color insert).

A.   User Interactions with Geographic Dynamics

Geographic dynamics can be enhanced, explored, and manipulated through user interactions with a GIS, including, for example, the new ideas of tangible user interface (TUI). A TUI integrates physical and digital representations of reality to enable the user to dynamically interact with familiar tangible objects (Ratti et al. 2004). Figure 4.1 shows a tangible worktable for the user to dynamically manipulate objects and terrain in the virtual environment. Recent attempts have been made to incorporate numerical modeling of geographic dynamics (such as erosion) in developing a tangible GIS to explore new directions to visualize geographic dynamics and human impacts on the environment (Mitasova et al. 2006).

B.   Visual Reality Models

Similar to the effects of dynamics generated by navigation, fly-through, or tangible user interactions, dynamic virtual reality models have been developed in the gaming industry with virtual reality, visual simulation, and 3D graphic techniques to handle scene transitions and interactions in forms of "story, art, and software" (Zyda 2005). The most important elements for success in video games are dynamically situated characters, complexity of landscape structures and views, as well as compelling and attractive story lines. These games represent dynamic storylines that occur in a virtual geographic domain.

Three general approaches are used to generate motion for characters in virtual environments: keyframing, motion capture, and dynamic simulation (Brogan et al. 1998). The keyframing approach is based on significant positions for the animated objects, and the algorithm interpolates intermediate frames between these positions. Motion capture starts with sensors to record movement according to shifting positions of body parts of human or animal subjects. The data recording position shifts are used to create graphic characters. In contrast, dynamic simulation applies the laws of physics to moving characters in a hierarchy of rigid body parts. Simulation of movements of any body part is computationally costly because calculations need to consider gravity, friction and collision, and torque at joints (for hopping and balancing, for example) to control behavior and locomotion.

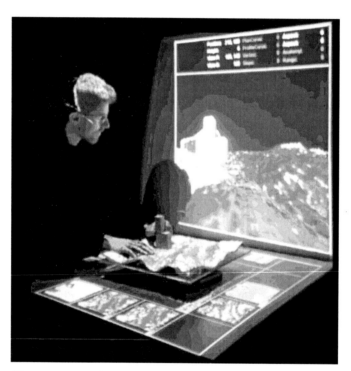

Figure 4.1: A tangible user interface that enables the user interacts with tangible objects in the virtual world. A vertical screen augments the experience with a virtual view of the landscape at human eye level (Ratti et al. 2004).

C.   Dynamics in Videos

Central to all the storytelling methods is to play out the historical or interactive courses of geographic dynamics. These narratives may be linear or nonlinear. When stories are too long or too complex and are beyond human cognitive capabilities to remember or decipher, some summative means would be of great value to replay and comprehend the stories. In video gaming, techniques are being developed to automatically generate movie summaries of interactions in video gaming through narrative extraction and movie construction based on log files that record actions taken by the player (Friedman et al. 2004).

Both linear and nonlinear narrative constructions need to convert the log file of actions and interactions into a data model that represents temporal sequence of actions (Figure 4.2). Each action item may include arguments like actors, positions, angles, and target objects. A bundle of actions with temporal or causal dependencies may be represented as routines (Figure 4.3). As a result, routines are higher level abstraction of actions that provide an effective data element to judge the degree of interest to be included in a movie summary.

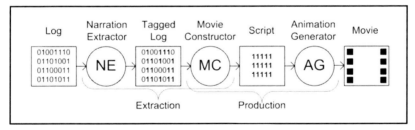

Figure 4.2: A workflow to construct movie summaries based on logged interactions during video gaming (Friedman et al. 2004). The figure aims to show the sequence of the procedures, i.e., log, narration extractor, tagged log, movie constructor, script, animation generator, and movie. The small characters are to show formats, not the context.

For movie, news, or other videos, summarization is important to provide highlights of stories as well as to search for scenes or interesting features. In general, video summarization techniques are categorized as static storyboard summary (i.e., keyframing) and dynamic video skimming (Ngo et al. 2005). Different from the keyframing approach, dynamic video skimming ensures temporal continuity through continuously browsing a video with highlights of important contents without sacrificing the temporal evolution of the video. In general, the dynamic video skimming and video zooming approaches consist of two steps: one to reduce spatial and temporal resolutions of video frames and, the other to develop tree or graph structures to hold temporal sequences of the low resolution frames (Figure 4.4).

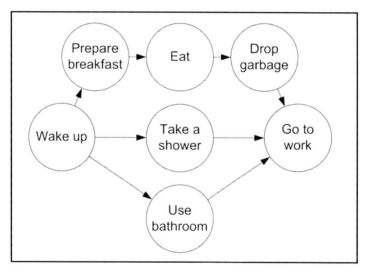

Figure 4.3: Examples of actions represented by circles in the diagram and routines represented by connected circles (Friedman et al. 2004).

An example of dynamic video skimming tools is the VideoZoom browser developed to provide effective and meaningful spatiotemporal video browsing on the Internet (Smith 1999). By alternatively zooming in space and time, the VideoZooming techniques start with an initial coarse view of the video and gradually drill down into different temporal parts of frame sequences, depending on relevancy to the content of interest in the video (Figure 4.5). As a progressive video retrieval approach, VideoZooming allows the user to identify the content of interest at a low resolution and retrieve the relevant frames in high resolution for detailed viewing. The video browsing and retrieving capability is afforded by spatial and temporal hierarchies of sequence compression in the video (Figure 4.6). Therefore, the technique is well suited for browsing large video files (such as NASA satellite images, sports videos, surveillance video, etc.) and retrieves the part of the video that is of interest to an investigation.

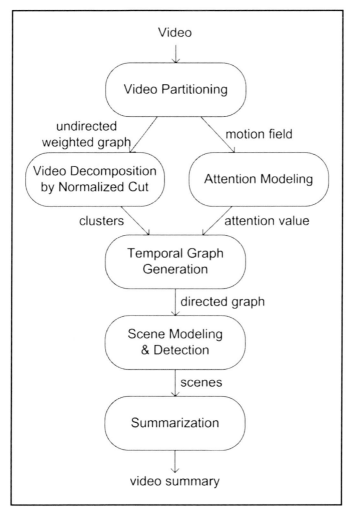

Figure 4.4: An example of dynamic video skimming procedures to create video summaries (Ngo et al. 2005).

The VideoZoom for video browsing is similar to VoxelViewer developed by Harrower (2002) to explore effects of the spatial and temporal granularity on patterns and feature extractions. Cartographic animation is, in essence, making cartographic videos with the objectives to reveal complex spatiotemporal patterns that are not evident in static maps. Time-series satellite images can be better understood through "storytelling" visualization to elicit spatiotemporal patterns and geographic change than a single "difference image" calculated by two images because the processes that drive the patterns and change may become apparent in an animation (Harrower 2002).

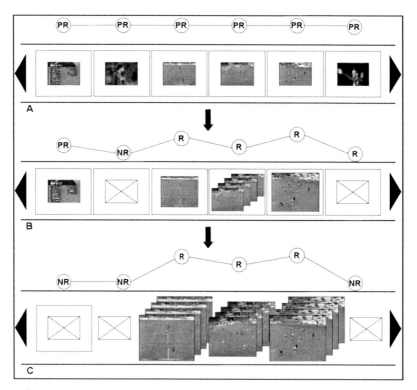

Figure 4.5: An example of progressive browsing and retrieving in a video. The procedures start with low resolution segments first and, at each stage of iteration, determine the relevancy of each segment (R: relevant; PR: probably relevant; NR: not relevant) to select segments for zooming farther in (Smith 1999).

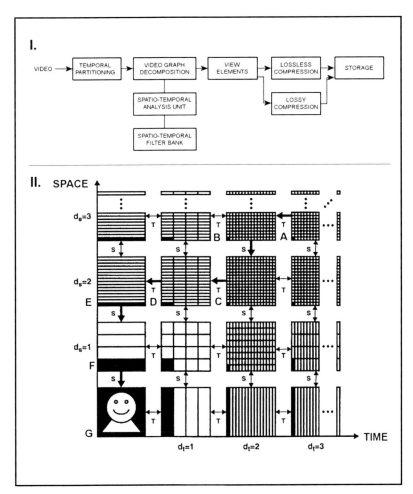

Figure 4.6: Spatiotemporal zooming approach to video summarization: panel I illustrates the conceptual model and workflows for the approach, and panel II illustrates an example of progressive video retrieval. The retrieval starts with the coarse view (A), zooms-in in time (B), zooms-in in space (C), zooms-in in time (D), zooms-in in space (E) and space again (F) to build-up the detail of the video sequence (Smith 1999).

In geography, change in space and time can be revealed in a time series of satellite images depending upon the spatial and temporal resolutions of these images. As posited in Hierarchy Theory processes operating at a larger scale are expected to exhibit lower frequency of variability in space and time (Ahl and Allen 1996). Changes driven by geographic processes require observations at different spatiotemporal resolutions. Studies of regional climate will need data at finer spatiotemporal resolutions than global climate. Determining the best spatial and temporal resolutions for change detection can be challenging (Figure 4.7).

Since 1990, technologies for geospatial data collection have grown exponentially with advanced sensor and survey systems. In addition to satellite remote sensing imagery, massive amounts of data are created at an ever-increasing pace through in situ environmental measurements, surveillance observations, transaction logs, individual reports, online postings, and many other means. Ironically, it is often found to be challenging to get the information that is needed from the practically boundless data amounts collected. The level of challenges is compounded when we attempt to identify and comprehend geographic dynamics in spatiotemporal data sets. Storytelling approaches to visualization take the first step by relating spatial data temporally, and are based on the temporal relations showing transitions and helping to recognize patterns and relationships in space and time.

Principles and guidelines for developing effective animated maps have been established in GIScience literature. In addition to the eight visual variables in the design of static maps, three dynamic variables (scene duration, rate of change between scenes, and scene order) are used to emphasize location, change, and attributes of the geographic phenomenon of interest (Dibiase et al. 1992). The effectiveness of methods and designs for map animation is evaluated by cognitive studies through experiments, interviews, and surveys. These cognitive experiments suggest four animation supplements to bridge perceptual biases in animated cartography: (1) the centrographic time-series map; (2) the biplot, a joint, two-dimensional representation of time units and places based upon two principal components; (3) canonical trend-surface analysis; and (4) the time-series correlation graph (Monmonier 1992).

As mentioned previously, vantage points of observation, user interactions, and simulation can also generate temporal relations. Cartographic animation and videos have been proven effective in telling stories about geographic dynamics. Similar to videos, cartographic animation would benefit from exploring summarization techniques for easy and quick browsing and retrieval of relevant and useful information. Many methods developed in video summarization should be potentially adaptable to cartographic animation. As in all storytelling, it is extremely useful to highlight trends, anomalies, or singular features from a temporal sequence of satellite images or climate change model outputs, for example. Such abilities are urgently needed to highlight relevant and interesting information and interactively select similar or critical features.

Figure 4.7: The spatial extent and latitudinal positions of its boundaries change seasonally and annually. Harrower (2002) suggested that the change can be best visualized in a 15 fps animation loop using 40-day averages with a 10Km pixel.

II.   VISUAL ANALYTICS

Visual analytics centers on using interactive visual interfaces to support analytical reasoning (Thomas and Cook 2005). While the term, *visual analytics*, has only been adopted recently (Wong and Thomas 2004), the concept that applies a visual means to analysis and reasoning has had a long tradition in a wide range of scientific practices. Statistical graphs and charts, for example, play major roles in effectively presenting and communicating results with visual summaries. Figures and diagrams provide "big pictures" of perhaps seemly disconnected concepts into an integrated view.

The old saying, "seeing is believing," concisely highlights the persuasive and convincing power of visual evidence. Nevertheless, visual analytics are more than providing visual means for displays. Equally important to visualization are the interactive capabilities provided with the visual means. Visual analytics involve intensive and extensive user interactions to manipulate the data, visual latent patterns, and synthesize deep information from massive data sets (Thomas and Cook 2005). The science of visual analytics covers four main areas of research (Thomas and Cook 2006):

- Analytical reasoning to discern deep insights for assessment, planning, and decision making

- Data representation and transformation to resolve conflicting and dynamic data for visualization and interaction

- Visual representation and interaction to enable exploring multi-dimensional data simultaneously

- Effective information production, presentation, dissemination, and communication of analytical results

The four main areas of visual analytics research together provide an end-to-end model for geographic dynamics understanding. Analytical reasoning aims at identifying contextual information and scenarios through understanding in order to develop measures for decision making. Interactive visual techniques enable spatiotemporal reasoning to be graphically represented and communicated in an integrated framework to illustrate multiple aspects of geographic dynamics across scales of space and time. Indeed, the four main areas of visual analytics research contribute a wide range of procedures in GIS processing and modeling (Figure 4.8). By displaying data in the form of maps and parallel coordinate plots, relationship and correlation become visually detectable to facilitate hypothesis generation. Interactive classification, charting, and mapping empower the user to freely explore what-if scenarios and examine results from different settings. Spatial overlays, data fusion, and modeling techniques afford inspections of meaningful information structure, potential causality and consequences of activities, events, and processes. Visualization also plays a critical role in presenting and communicating results and eventually tells the stories of what has happened, how and why it happened, and what may be next.

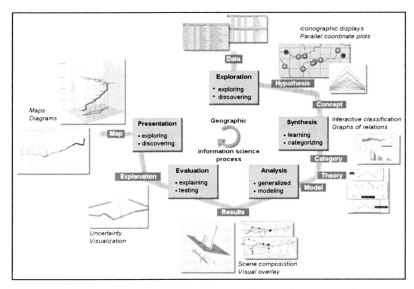

Figure 4.8: A wide range of visual techniques used in GIScience research (Kraak 2004).

A.   Visual Analytical Reasoning

A common example of visual analytical reasoning is the use of Venn diagrams to show relationships between sets, such as intersections and unions in polygon overlays. Sowa's Conceptual Graph Theory (Sowa 1984) also provides an effective visual means to represent knowledge in contextual structures that is logically precise, humanly readable, and computationally tractable. Figure 4.9 shows a conceptual graph with four concepts: person, go, city, and bus; and three relations: agent, destination, and instrument (Sowa 2000). The graph represents a fact that John goes to Boston by bus. By introducing temporal intervals, temporal objects, temporal situation, and temporal relation, conceptual graph theory can be extended to represent and inference temporal knowledge (Moulin 1997), as shown in Figure 4.10. All three panels represent three levels of temporal discourse about "John goes to Toronto" on communication (Panel A), time line (Panel B), and an extended conceptual graph (Panel C). While the three panels convey similar temporal discourse that John goes to Toronto, each visualization means has its own strength in expressing temporal discourse.

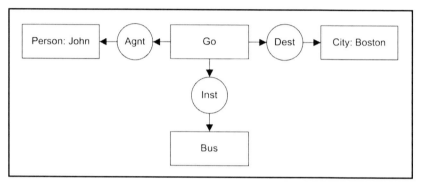

Figure 4.9: An example of a conceptual graph with four concepts and three relations (from Sowa 2000 http://www.jfsowa.com/cg/cgexamp.htm#Ex_3).

In education, for example, it has been shown that concept maps effectively support visual learning (Nesbit and Adesope 2006) and tools for constructing concept maps such as Cmap[2] have been applied effectively for querying and mining on the Web (Cañas et al. 2004a; Cañas et al. 2004b). Three essential elements of a concept map are concepts, propositions, and hierarchical structures. By relating the three elements on a subject, construction of a concept map improves knowledge understanding and retention. Not only benefiting the visual learning of individuals, concept maps also facilitate collaboration on tangible and abstract subjects through group efforts in building analytical constructs of contextualization for individual concepts and their relationships.

Visualization through concept mapping promotes the processes of distilling key conceptual elements, their relationships, and the hierarchical structure that ties together these concepts and relations, and, therefore, facilitates learning and understanding. However, there is a learning difference between viewing and visualizing concept maps; experimental research showed that a visualization group involving activities to generate concept

---

[2] Cmap is available at http://cmap.ihmc.us/

maps outperformed a viewing group with activities only to inspect existing concepts maps (Lee and Nelson 2005). While the experiment only confirmed the effect on well-structured problem-solving performance, the outcome suggests the importance of interactive components in visual analytics.

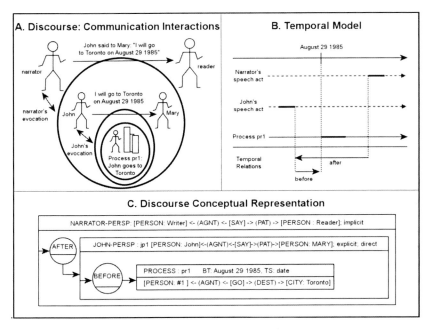

Figure 4.10: Three levels of temporal discourses (Moulin 1997). With kind permission from Springer Science and Business Media.

The GIScience community has also applied visual analytical reasoning successfully, for example, for representing categories of relations, most notably for topological relations, where distinctions between relations that hold for region objects and region-line combinations are visualized (Egenhofer and Franzosa 1991; Randell et al. 1992; Mark and Egenhofer 1994). In other work, a visual iconic language referred to as the change description language (Hornsby and Egenhofer 2000) is another example of visual analytical reasoning. In this case, changes to identifiable objects (e.g., nation states) over time are modeled through an iconic representation that models and tracks the changes to object identities over time.

B.   Data Representation and Transformation

The second broad topical area of visual analytics science is on data representation and transformation with aims to ease reasoning, data analysis and modeling and support data mining and knowledge discovery. Two dimensional and three dimensional presentations can provide effective frameworks to discern spatiotemporal patterns and relationships. Many multi-dimensional scaling methods are available to reduce data dimensionality to visual forms. The commonly used principle component analysis, factor analysis, and multi-scaling methods can transform data into a manageable number of dimensions to facilitate recognition of meanings deeply embedded in complex data sets.

Novel visual approaches to support knowledge construction are being developed for new data representation methods or new data transformation methods. In general, there are two approaches taken in the development of visual techniques: data representation methods and other methods applied to these representations, such as projection techniques (Gahegan et al. 2001). Here, we categorize the latter approach as data transformation methods. Data representation methods include a wide range of map-based techniques, chart-based techniques (such as scatter plots or parallel coordinate plots), and hierarchical and network techniques to organize data effectively in a particular structure, such as a tree. Data transformation methods, on the other hand, include projection techniques based on statistical data transformations, pixel techniques to create visible clusters, and iconographic techniques to display multifaceted data attributes simultaneously. Gehegan et. al (2001) offer an excellent overview of these techniques and visual support for data exploration, knowledge construction, and geocomputation. While map-based and chart-based techniques are common in GIScience research, data transformation methods have attained recognition in their visual support for spatial and temporal analysis.

Among all data transformation methods, Self Organizing Maps (SOM) have recently emerged as a popular tool for data transformation (Kohonen 2001). SOM algorithms transform high-dimensional data into two-dimensional space based on vector quantization of distance measures to reveal quantitative interrelationships among data records in a predefined grid of rectangular, hexagonal, or other kinds of basic spatial units. By placing similar data records (i.e., shorter distance measure in vector quantization) in closer locations on a grid, the data records are transformed into a spatial form in which spatial patterns of similarity among these data records can become apparent. As spatial metaphors are effective ways to reveal patterns and relationships, user experiments show that data records falling in the same region on a SOM are perceived to be more similar than those in different regions (Fabrikant et al. 2006). The hierarchical nature of spatial associations applicable to both maps and attribute space suggests when transformed data representation to visual forms, mapping techniques (such as color selection, symbolization, and visual variables) can be used to effectively represent and assess attribute similarity and structure of data records.

In addition to data dimension reduction, data representation and transformation algorithms address issues of uncertainty, data fusion, and both scaling and dynamic natures of geographic data (MacEachren et al. 2005). These tasks are central to data mining and knowledge discovery. Different visual forms of data representation vary in their effectiveness to facilitate understanding of spatial relationships and processes (Andrienko et al. 2002; Koua et al. 2006). To assess use and usability of geovisualization techniques for data exploration, Koua and colleagues transformed a data set to four different forms of visual representation to compare their effectiveness in acquiring information about relationships between geography and macroeconomic growth. The data set consisted of 48 variables on economy, physical geography, population, and health for 150 countries worldwide. Their result suggested that the comparative effectiveness of maps, parallel coordinate plots (PCP), SOM distance matrices, SOM 2D/3D surfaces, and SOM component plane displays vary with each task.

SOM-based visualization was effective to facilitate tasks on grouping, clustering, attribution correlation, and SOM component plane displays are particularly effective for revealing patterns in the data and relationships among attributes. As expected, maps are in general more effective than the other visual forms for tasks that involve locating, distinguishing, and ranking. Their finding echoes the need for an approach that combines multiple forms of visualization to meet a wide range of visual needs for understanding and knowledge discovery (Guo et al. 2005).

Another layer of challenges for applying visual analytics of geographic dynamics is the need to cover both spatial and temporal dimensions with equal importance. Instead of considering spatial and temporal variances separately, geographic dynamics understanding requires a joint examination of spatiotemporal variations of multivariate patterns. For example, a Visualization System for Space-Time And Multivariate Patterns (VIS-STAMP) was recently developed to meet the challenge of spatiotemporal visual analytics (Guo et al. 2006). VIS-STAMP combines cartographical, visual, and computational methods to assist exploring and cross-referencing spatial, temporal, and multivariate patterns simultaneously. Its underlying strategy to enhance effectiveness in visualization is through data abstraction, as opposed to resolving data overlaps by sampling, density mapping, shifting data points, or extending data space with the Wide Area Layout Data Observer (WALDO) tools (Keim et al. 2004). VIS-STAMP applies SOM, PCP, and two reorderable matrices. One of the matrices is used to organize multivariate patterns, and the other is able toreveal spatial variation of multivariate patterns (see example below).

Using 1992 to 2003 company data from sixteen industry types across forty-eight states and Washington, D.C., Guo and his colleagues demonstrated the support of VIS-STAMP to elicit spatiotemporal patterns and relationships by reorderable matrices, maps, SOMs, and PCPs with the same color scheme across all these visual representations where PCPs also serve as legends. The reorderable matrices in Figure 4.11 illustrate temporal variations of company data for each industry type. States are grouped into regions (e.g., Pacific: Pac; Southwest: SW, etc.), and each reorderable matrix highlights industrial types: their spatial distribution (columns) across the respective region and their temporal distribution (row) within each state. Here, different colors represent different industry types. Each dot in the SOM represents a cluster of an industry type, and the dot size is in proportion to its sales value. A time series of maps directly display spatial patterns of these industries year to year and how the pattern has transitioned from 1992 to 2003. The PCP provides an effective visual cue to highlight the industry types with high and low sales values over these years.

Alternatively, the reorderable matrix and maps in Figure 4.12 emphasize the spatial distribution of each industrial type. Nevertheless, the reorderable matrices show what industry types are dominant within a given state, whereas, the maps show how these industry types are distributed across the continental United States. The SOM shows how these industry types may cluster in the attribute space of sale vales. Temporal variation of sales values within each industry type is provided in the PCP. In addition, the PCP also indicates that different colors here represent different levels of sales values (with white being of low sales values). By the same token, all the white states in the maps and the white cells in the reorderable matrix represent low sales values for a given industry type.

These examples of the data representation and transformation aspects of visual analytics clearly communicate the importance of offering multiple ways to visualize spatial and temporal data for improved understanding. Complementary to the visual analytical reasoning supported by conceptual graphs, workflow diagrams, or concept mapping, data representation and transformation dig into data mass to reveal embedded deep information that can be used to further refine and enrich visual analytical reasoning. In addition, user interaction is critical to effective data exploration (Guo et al. 2003). Abilities to browse, brush, and change classification schemes, for example, can lead to discovering new patterns and relationships, and furthermore revealing possibilities for novel alternatives to problem solving.

The capability of visual representation and interaction to enable exploring multi-dimensional data simultaneously empowers the user to draw new insights into change and movements resulted from activities, events, and processes. Such insights are invaluable to developing hypotheses, formulating reasoning and modeling schemes, and eventually constructing new knowledge about geographic dynamics. Visualization, in the end of the understanding and knowledge construction processes, can further serve as an effective tool for information production, presentation, dissemination, and communication of analytical results, which may further stimulate new thoughts, research questions, and hypotheses for yet another new cycle of understanding and knowledge discovery of geographic dynamics.

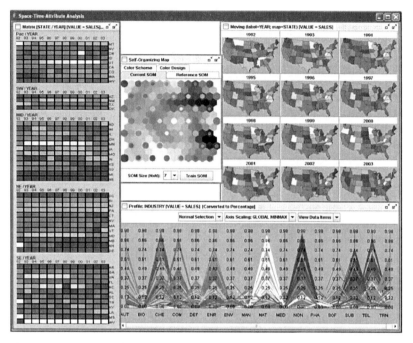

Figure 4.11: (See color insert.) VIS-STAMP visual displays with an emphasis on temporal variation of sales values across industrial types and across the continental United States. The SOM and PCP complement with distributions in the attribute space of sales values (Guo et al. 2006).

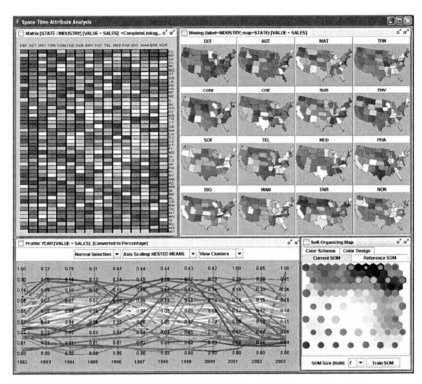

Figure 4.12: (See color insert.) VIS-STAMP with emphases on spatial distribution of each industrial type across forty-eight states. PCP complements the temporal dimension of visualization by displaying the trends of these industry types from 1992 to 2003. The SOM is the same as the SOM in Figure 4.11 (Guo et al. 2006).

CHAPTER 5

ANALYSIS AND MODELING APPROACHES FOR GEOGRAPHIC DYNAMICS UNDERSTANDING

- *What are different ways to visualize geographic dynamics?*
- *How can different aspects of geographic dynamics be visualized and understood?*

Space and time are essential to analysis and modeling of geographic dynamics. Activities, events, or processes, as well as change and movement, are all considered as functions of space and time. A mathematical or statistical model for geographic dynamics, therefore, consists of spatial and temporal variables. These spatial or temporal variables can be explicit or implicit. Explicit variables directly reference spatial measurements (such as distance or coordinate pairs for locations) or temporal measurements (such as a particular instance, e.g., 10 a.m. or interval, e.g., every ten hours). For implicit variables, space and time references are hidden in forms of other measures, such as speed of movement or rate of change in area.

Selection of spatial and temporal variables in model development depends upon the nature of geographic dynamics to be modeled. Spatial and temporal explicit variables apply discrete measures to geographic dynamics, even though the geographic dynamics of interest exhibits properties of continuity. For example, a space-time path of an activity (e.g., a trip to a school, a mall, and a coffee shop, and then back home) may be continuous, but the observations or records of the activity (e.g., geospatial coordinates of the school and arriving time at the school) are taken in a discrete fashion. By contrast, spatiotemporally implicit variables capture well continuous properties in geographic dynamics. An example of spatial and temporal implicit variables used in modeling geographic dynamics is slope modeling. A model of hill slope development will place emphases on erosion and deposition rates resulting from surface runoff, undercutting, sediment transport, flow, and other factors (Martin 2000; Mills and Mills 2001). Certainly, a model can include both explicit and implicit variables of space and time to address the needs for both discrete and continuous measures of geographic dynamics.

As discussed in the previous chapter, visualization approaches can be categorized as storytelling or visual analytics. Analysis and modeling of geographic dynamics appear to share two similar tracks with, as we name them here, descriptive and predictive approaches. The distinction can be fuzzy. Descriptive approaches can support prediction by projecting the unveiled stories into the future. Predictive approaches also depend upon the understanding of spatiotemporal patterns that summarize trends and possibilities to make predictions. For the purpose of the following discussions, the distinction is to distinguish emphases taken by the two approaches—one on what has happened, and the other on what is likely to happen.

I.   DESCRIPTIVE APPROACHES

Descriptive approaches to the analysis and modeling of geographic dynamics tell the story of how drivers of geographic dynamics have provoked change and movement in space and time and how to identify areas of time of interest with a high or low intensity of geographic dynamics.

Common descriptive analysis and modeling techniques include density analysis, hot spot detection, cluster analysis, and spatiotemporal interpolation. Given a spatiotemporal dataset of crimes, for example, descriptive modeling aims to discover areas of hot spot and to reveal change in crime patterns over time. By monitoring crime patterns, descriptive modeling can suggest periods and areas of high crimes and of rising or falling crime density (Ratcliffe and McCullagh 1998). Routine Activity Theory posits that most crime requires the spatiotemporal convergence of a motivated offender, a suitable target, and the absence of a capable guardian (Cohen and Felson 1979). Such spatiotemporal convergence can be related to other changes in other geographic variables (e.g., income, ethnicity, infrastructure, etc.) for interpretation to reveal potential solutions (Pain et al. 2006). While there are societal concerns about crime mapping (Ratcliffe 2002), spatial and temporal analysis of crimes through GIS technology has become popular in police practice and research to understand spatiotemporal dynamics of crimes (Bowers et al. 2004; Weisburd and Lum 2005).

Descriptive analysis and modeling of crimes, in general, consider crimes as point-based events in space and time. The same conceptualization is applicable to other events, such as traffic accidents, disease infection cases, the presences of a certain species, tornado climatology, terror attacks, and many other events. For these cases, the data mainly denotes the time and location of occurrences and the principal concern is on distribution patterns in space and time. There are many descriptive modeling methods and tools for spatiotemporal point pattern analysis. STAC[1] (Spatial and Temporal Analysis of Crime), CrimeStat,[2] TerraSeer,[3] Splancs[4] (Spatial Point-Pattern Analysis Code in S-plus and Maptime (Slocum et al. 2000) are just a few examples. STAC was developed with support from the U.S. Department of Justice and U.S. Census Bureau. The software has been widely used by police agencies in detecting crime hotspot areas since 1996. STAC identifies crime hot spots by iteratively searching for the densest ellipse and is now part of the CrimeStat[5] program (Levine 2006) that provides comprehensive functions for mapping, analysis, modeling, and visualization of spatiotemporal crime data (Figures 5.1 and 5.2). While the descriptive approach is based upon the spatiotemporal data of events that tell stories of event distributions and pattern changes, the revealed "stories" can be useful for prediction (Figure 5.3).

In addition to point-based analysis and modeling, STARS (Space-Time Analysis of Regional System) consists of spatiotemporal analysis tools for spatially extensive data, i.e., polygonal data or data with spatial extent (Rey and Janikas 2006). With both visualization and geocomputational modules, STARS incorporates tools to perform descriptive statistics

---

[1] STAC is available free at http://www.icjia.state.il.us
[2] CrimeStat is available free at http://www.ojp.usdoj.gov/nij/maps
[3] TerraSeer is a proprietary software package. More information is available at http://www.terraseer.com/
[4] Splancs is available free at http://www.maths.lancs.ac.uk/Software/Splancs/

(distribution and summary measures), exploratory spatial data analysis (Moran's I, Geary's c, LISA, and G statistics), inequality (Gini coefficient, Theil decomposition), mobility (τ and θ statistics), and Markov analysis (classic and spatial Markov techniques). A suite of novel spatiotemporal mapping and graphing tools in STARS empowers the user to identify similar change paths and detect co-movement dynamics. However, STARS is limited to analysis of attribute data of defined areas, and, thus, there is no tool to analyze changes that occur to geometry or topology.

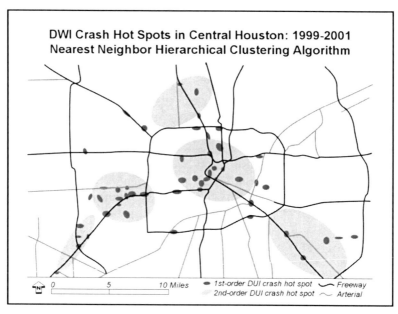

Figure 5.1: Identification of crime hotspots (Levine 2006).

To account for changes to geometry and topology, ideas are proposed and tested by GIScience researchers. For example, Sadahiro (2001) provides an exploratory method with qualitative directed graphs and quantitative metrics to access changes to geometry and topology among unmovable polygons and tested his methods to analyze changes of market areas of a convenience store chain in Tokyo, Japan. For polygons representing moving objects, directed graphs are also experimented to track geometrical and topological changes within and across polygons of wildfire burns to record the progression of a fire (Yuan 1999).

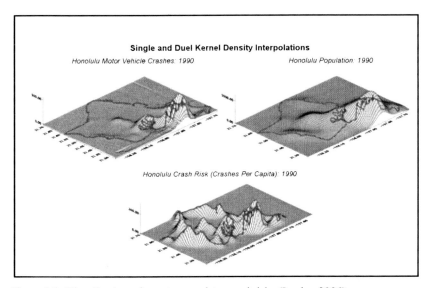

Figure 5.2: Visualization of events, correlates, and risks (Levine 2006).

In addition to point-based events, epidemiologists and political scientists consider diffusions and lifelines of events. Modeling historical events has emphasized both the duration of the initial state and the transition to subsequent states. For social scientists, event history data includes the time of the units beginning the process under study and the time of event occurrence (Box-Steffensmeier and Jones 2004). Mathematically, event history analysis starts with defining a positive random variable of survival time and examines the probability of the survival time under the probability of an event that may occur at an instantaneous time to terminate the survival time. By defining the survival function and the density of failure times in which an event ends a state and transitions to a subsequent state, social scientists gain insights into the survival and death accounted for in the event history framework. The relationship between failure time and the survival function reflects the hazard rate at which a state fails (or duration ends) by some time *(t)* given that the state survives until that time *(t)*. While the event history modeling may include geographic entities, such as countries or states, spatial locations or properties are not considered in the formulation of a survival function or the density of failure times. A potential possibility to account for spatial effects is to incorporate neighborhood analysis, such as spatial autocorrelation or diffusion modeling, in event history analysis.

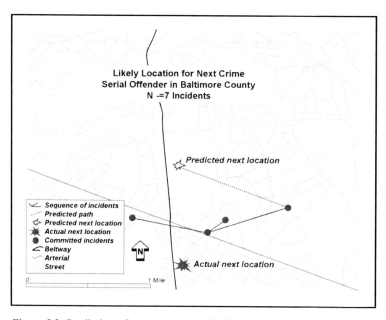

Figure 5.3: Prediction of events (Levine 2006).

Diffusion modeling is one key element in spatial epidemics, while some spatial epidemiological studies incorporate spatial considerations through considerations of continuous or discrete distributions in space. Spatial epidemics applies both deterministic and stochastic approaches with the contact distribution concept to model the spread of infection (Rass and Radcliffe 2003). Deterministic approaches, in general, assume a homogeneous population and when an individual becomes infected, the individual cannot return to the susceptible state. Under the assumption, the spread of infection is modeled as a wave traveling through the population.

Based on weekly measles reports in London from 1944 to 1994, Grenfell et al. (2001) constructed time-average phase maps to identify spatial hierarchies of waves that move from large population centers to surrounding rural areas. The year 1968 marked the prevaccine and postvaccine periods. They focused in the period of 1950 to1966 for the analysis of spatial dynamics. The authors concluded that "the dominant waves are around London and Manchester: the major epidemic occurred 4-6 weeks on average before London in the urban northwest, moving from Liverpool and Manchester into the north Midlands" (Figure 5.4).

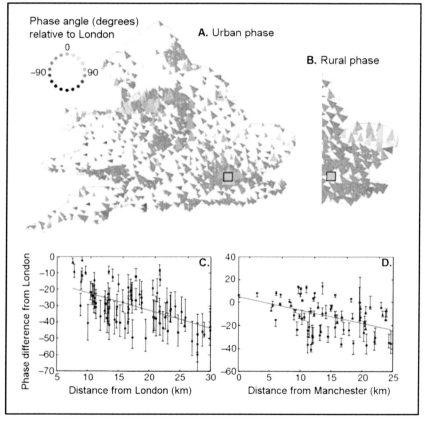

Figure 5.4: Phase differences analysis of measles epidemics in London and surrounding areas (Grenfell et al. 2001). Reprinted by permission from Macmillan Publishers Ltd.

Furthermore, spatial dynamics of epidemics can be analyzed through tracking changes to a spatial wave over time and identifying the leading and trailing edges of advances and retreats of the wave (Cliff and Haggett 2006). Figure 5.5 shows that the conceptual model depicts the phases of an introduction of an infection to an island represented by the shaded area. Central to the analysis is a space-time data matrix that summarizes the time of reporting (columns), the area of reporting (rows), and sums of infection cases for each area over time, and sums of infection cases for each time over the areas. By a space-time data matrix and net spatial change in infection cases, the leading edge and following edge of the infection wave can be identified (Figure 5.6).

Deterministic spatial epidemics are based on continuous space point processes (Barlett 1956), assuming a contact distribution among stationary individuals in a homogeneous population. Stochastic spatial epidemics, on the other hand, centers upon the concept of distributed infectives that epidemics are spread by the diffusion of infected individuals (Reluga et al. 2006). Recent work in spatial stochastic epidemics modeling is mainly based on discrete space models to account for spatial heterogeneity of population, population susceptibility, and patchiness in the spread of the infection. By using spatially explicit models, spatial epidemics modeling can examine the effects of spatial heterogeneity on epidemic outbreaks (Sander et al. 2002).

Spatial heterogeneity and spatial scale, furthermore, may suggest multiple levels of the spread of a disease. Dengue epidemics, for example, are found to be related to spatial heterogeneity at two spatial scales of contacts: a local level with contacts within households, and a global level with contacts between people from different households (Favier et al. 2005). Simulation of stochastic spatial processes is supported by several simulation or geocomputational methods. The need to generate a large number of realizations to estimate behaviors of a spatial epidemics system is computationally demanding. Alternatively, approximation techniques are used to develop simpler, analytical methods as surrogates for predictive tools, such as cluster approximations to studying models with nearest-neighbor interactions (Favier et al. 2005). The Markov Chain Monte Carlo modeling, cellular automata, agent-based modeling, and spatial evolutionary modeling all support stochastic spatial epidemics and other spatially explicit dynamics in geographic domains. These methods provide potential scenarios in the development of geographic dynamics, and, therefore, align well with the emphasis of predictive modeling methods for geographic dynamics.

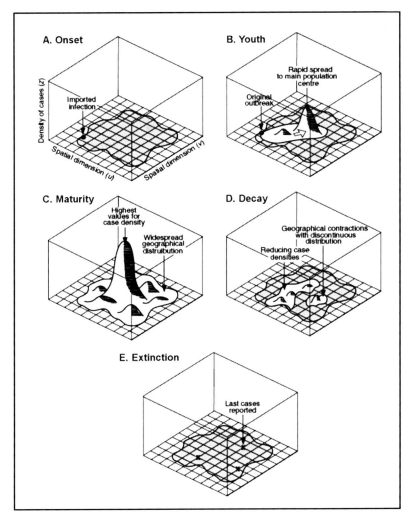

Figure 5.5: A conceptual model of tracking a spatial infection wave (Cliff and Haggett 2006). With kind permission from Springer Science and Business Media.

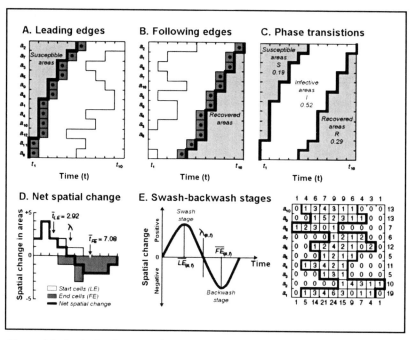

Figure 5.6: Analysis of a space-time matrix (lower right) can reveal the leading edge (A), following edge (B) and phase transition (C). The net spatial change analysis (D) can be generated to a swash-backwash model (E) in a continuous space (Cliff and Haggett 2006). With kind permission from Springer Science and Business Media.

II.   PREDICTIVE MODELING METHODS FOR GEOGRAPHIC DYNAMICS

Predictive modeling approaches to geographic dynamics center on simulation or geocomputation. The term, "geocomputation," was introduced in 1996 at the first International Conference on GeoComputation at the University of

Leeds.[6] Aiming at developing computational methods for geographic analysis, geocomputation encompasses research efforts to develop analytical and modeling tools for nondeterministic, complex geographic problems (Gahegan 1999). While geocomputation studies may or may not consider temporal variables or dynamics, many geocomputation researchers focus on topics, such as urban sprawl and land use and land cover change (LULC) that are innately spatiotemporal. Geocomputation is an active and growing field of research among quantitative geographers, GIScientists, and computer scientists, even though it has been only eleven years since its introduction from the academia. A comprehensive review is beyond the scope of this workshop report. Interested readers are recommended to consult a special issue of *Transactions in GIS* on Geocomputation (volume 3, issue 3, published in 1999). The discussion that follows highlights the relevance of geocomputation to geographic dynamics. In particular, discussions will cover key ideas in Markov Chain Monte Carlo (MCMC) methods, cellular automata (CA), agent-based modeling (ABM), and evolutionary modeling or genetic algorithms (GA).

MCMC methods combine Markov Chain techniques to model state transitions and Monte Carlo methods to simulate randomness in each state by defining some probability density functions. Specifically, Markov Chain methods are used to generate samples from a predefined transition probability function. The samples are a series of simulated invariant states to a defined target probability. A first-order MC generates a new state with the immediately previous state and the transition probability function. A higher order Markov Chain method considers second order or more distant states to the state to be generated. Monte Carlo methods, on the other hand, are stochastic statistics that draw random numbers (or random walk) from uniform probability distributions.

When Markov Chain and Monte Carlo methods are combined, random samples (or random walkers) move according to the defined probability distribution as specified by the Monte Carlo method, and then the Markov Chain method takes the integral from all walkers and the defined

---

[6] Since 1996, the International Conference of Geocomputation was held annually until 2000. Since then the conference has been held every other year since 2001. The conference Web site (http://www. geocomputation.org) archives extended abstracts of papers presented at all GeoComputation conferences.

transition function to determine the next state. Computationally, the MCMC approach is not trivial because of the large number of possible walks and possible states. Nevertheless, the MCMC approach is a general purpose computing technique with broad applications in biomedical, epidemiological (O'Neill et al. 2000; Nobre et al. 2005), biological (Fonnesbeck and Conroy 2004), sociological (Lewis and Raftery 1999), economical (Verhofen 2005), and climate (Tomassini et al. 2007) simulations as well as image processing in remote sensing (Haario et al. 2004) and analysis of uncertainty (Godsill 2001).

As an illustration, MCMC methods with Bayesian modeling are used to analyze nonstationary spatial and temporal data for environmental processes, such as pollution and spread of disease in three stages (Wikle et al. 1998). First is to specify a measurement process (or referred to as state process) in observation data. The measurement process represents the process of interest to the observer who attempts to extract from the measurements with the understanding that the measurements consist of errors. The second stage is to determine site-specific time series models for the state variable considering large-scale variability and a space-time dynamic process for the anomalies. The state process of interest (Y) may display multi-scalar variability and can be modeled by sub-processes (X_i) and a collection of parameters. In atmospheric science, Y may be an annual climate process, and Xs represent weather processes or seasonal processes that Y conditionally depends upon.

Next, the third stage needs to express spatial structures and dynamics by constructing Markov random fields that represent the parameters of the spatially dependent time series models. Finally, the fourth stage aims to specify priors or hyperpriors (if any) for all model parameters. Parameterization depends on application domains to determine proper conditional or empirical estimates. Bayesian techniques are applied to estimate prior probability distributions for parameterization. The dynamic system is then implemented in a MCMC framework with Monte Carlo methods to simulate errors embedded in the state process and building Markov random fields, and then Markov Chain transition functions are applied to model state changes over time.

MCMC methods incorporate random walks or uniform probabilities to state processes, while cellular automata (CA) methods apply rules to guide state transitions. Agent-based modeling (ABM) adds mobility to cells for mobile objects that are capable of assessing conditions and making decisions accordingly. Another dimension of complexity is introduced by genetic algorithms for Spatial Evolutionary Modeling (SEM) that emphasizes learning, reproduction capability, cross-fertilization, and mutation of individual objects. CA, ABM, and SEM are spatially disaggregated techniques capable of modeling individuals' local decisions to simulate global dynamics. Complexity Theory offers the theoretical underpinning for the use of these disaggregated approaches in order to address behavior and responses of a spatial complex system. Nevertheless, the distinction between CA and ABM may not be agreeable to all researchers on the subjects. The discussions below mainly follow a series of papers published by Michael Batty and his colleagues as summarized in (Batty 2005).

In CA, a cell changes its state based on state transition rules that usually account for local neighborhood effects. There are generally limited and often categorical states possible for a given cell. Transitions from one state to another depend upon a set of preset rules that guide individual cells to react, diffuse, or randomly change. CA techniques have been broadly used in urban land use and land cover change (Dietzel and Clarke 2006), urban growth modeling (Liu and Phinn 2003; Torrens and Benenson 2005), wildfire modeling (Millington et al. 2006), spread of diseases (Bone et al. 2006), and many other application domains in which change of land surface characteristics is important.

CA approaches enable representation of spatial heterogeneity in modeling of complex systems. Ecological modeling, for example, is able to incorporate diverse environmental components and biological variables in a multi-scalar, nonlinear framework to model dynamic landscape ecological systems (Wu and David 2002) and to investigate edge effects (Kupfer and Runkle 2003) and competition among species (Chen et al. 2002). While CA conventionally applies grid-based representations of space, networks or point-based representations of space can also work with CA algorithms (Bonacich 2003; Bithell and Macmillan 2007).

While CA emphasizes state change at locations, ABM incorporates the mobility to model agents walking in space. An ABM simulation can start walkers in random locations. If the random process is considered as the driver of geographic dynamics of agents' distribution, random walk can be assigned to represent how these agents explore the modeling space. Alternatively, goal-seeking activities can be specified to guide walkers' movements. Several ABM systems have been developed to simulate pedestrian movements in urban environments. The STREETS model uses socioeconomic and other geospatial data to establish a statistically reasonable population of pedestrians and then simulates pedestrian movements under the constraints of spatial configuration, predetermined activity schedules, and land use distribution (Haklay et al. 2001). Figure 5.7 illustrates the design and structure of the STREETS with a GIS environment that represents physical and social constructs of a closed city system and a Swarm environment that model the pedestrian population and moving behavior. The movement of an agent is constrained by distribution of walkability in the city environment, maximum walking speed, visual range of the agent's visual acuity, and preset activities.

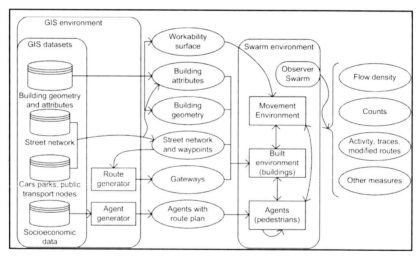

Figure 5.7: The design and structure of the STREETS model (Haklay et al. 2001). Pion Limited, London.

Another ABM system, PEDFLOW, simulates behavior and conflicts at a micro scale where individual pedestrians respond to each other and to personal space and awareness (Kerridge et al. 2001). Based on previous studies, Kerridge and colleagues incorporated the maximum rate (27 persons per foot width per minute at a concentration of 0.11 persons per ft^2) to model pedestrian flows in subways. PEDFLOW determined distance of movement for each individual through minimal assumption of pedestrian behavior. A similar approach is taken to examine emergent path patterns by individual pedestrians' preferred walking behavior and behavioral change by social fields or social forces (Helbing et al. 2001). For example, crowds moving towards opposite directions tend to form streams of pedestrians with a self-organizing attempt to move in the same direction. The formation of the concerted streams expedites the pedestrian flows and reduces probability of collision (Figure 5.8).

ABM empowers researchers to model individuals' behavior and decision-making processing with which spatial and temporal dynamics at a global scale may emerge to reveal the complexity of a spatial system. Of course, agents in ABM can be other than persons. Individuals can be animals (Bennett and Tang 2006), diseases (Gordon et al. 2003), or land use and land cover units (Parker et al. 2003), to name a few.

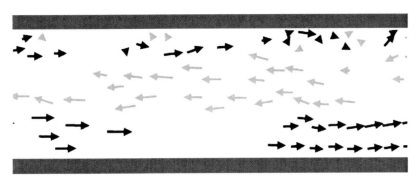

Figure 5.8: Self-organizing streams of pedestrian movements at opposite directions (Helbing et al. 2001). Pion Limited, London.

While individuals in ABM can be equipped with learning capabilities, ABM approaches only consider the modeled individuals independent of biological ties to others. The rising popularity of genetic algorithms and artificial life simulations facilitates the development of spatial evolutionary modeling that accounts for biological implications and ecological circles of life among individuals of multiple generations in a complex adaptive spatial system. Every individual in spatial evolutionary modeling is associated with a genetic combination and is subject to changes and transformations. A spatial evolutionary model (SEM) concerns not only responses from individuals, but adjustments from a population. Basic terms used in SEM are from evolutionary biology (Krzanowski and Raper 2001):

- Individual: an evolutionary unit

- Gene: a section of a chromosome

- Population: a set of individuals

- Fitness: a measure of an individual with respect to the objectives of evolution

- Objective functions: mapping individuals' fitness values to the problem space

- Generation: one evolution cycle

- Parent population: individuals in the previous generation

- Offspring population: individuals in the next generation

- Evolutionary operators (including cross-over, mutation, and selection): transformations carried out by individuals during an evolution

- Initiation of a population: a process to generate an initial set of individuals

- Terminating functions: define conditions that stop evolution

Figure 5.9 illustrates the key ideas of population, gene, features, chromosomes, and individuals (A) and concepts of cross-over (B) and mutation (C). A typical genetic algorithm builds upon these concepts to simulate biological evolution (Figure 5.10). The algorithm starts with an initial population. For each individual in the population, a fitness value will be calculated and assigned based on a predefined fitness function. If termination criteria are met, the evolution stops; otherwise, the algorithm proceeds to select individuals to mate and cross-over their chromosomes in their new generation. Mutation operators can randomly intervene during the evolution process and cause random changes in chromosomes and, thus, result in a new fitness calculation.

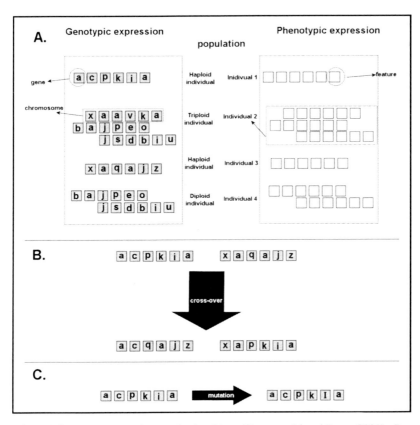

Figure 5.9: Key concepts in genetic algorithms (Krzanowski and Raper 2001). By permission of Oxford University Press.

Genetic algorithms have been used in GIS for knowledge classification (Armstrong 1991), network analysis (Wu 2000), forest management (Ducheyne et al. 2006), hazard transport planning (Huang et al. 2004), and many other semi-structured spatial problems to which researchers seek optimal solutions. Genetic algorithms are well suited for solving optimization problems, such as label placement (Van Dijk et al. 2002), spatial search (Li and Yeh 2005), land use planning (Matthews et al. 2006), and others.

While genetic algorithms or spatial evolutionary modeling are, by definition, dynamic applications of spatial evolutionary modeling to address geographic dynamics are lacking. There are opportunities that combine agent-based modeling and spatial evolutionary modeling by enabling individual agents to mate, reproduce, and mutate in a spatial complex system. The combination of cellular automata, agent-based modeling, and spatial evolutionary modeling will improve understanding of geographic dynamics to consider variability and adaptability of the mechanisms, functions, development of dynamics for the environment, the individuals, the populations, and their interactions. By doing so, researchers can better address multifaceted, multi-scalar, and multi-dimensional drivers of geographic dynamics (e.g., activities, processes, and events) and spatiotemporal consequences (e.g., change and movement) resulting from the drivers.

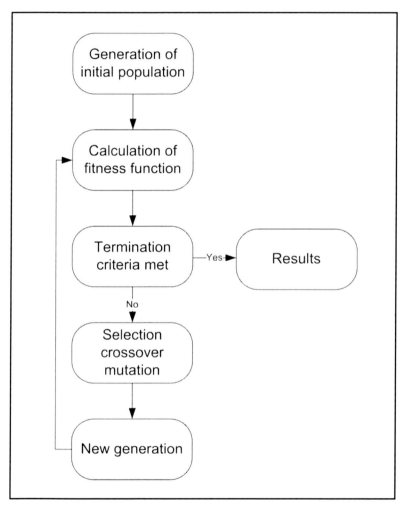

Figure 5.10: A simple evolutionary algorithm (Kumar et al. 2007). With kind permission from Springer Science and Business Media.

CHAPTER 6

RESEARCH CHALLENGES AND RECOMMENDATIONS

- *What are open research challenges relating to dynamics?*
- *What are the long-term versus short-term challenges?*
- *What recommendations can encourage further collaboration and communication?*

A goal of the workshop was to define a research agenda relating to the area of computation and visualization for understanding of dynamic geographic domains. Participants were asked to identify what they saw as major research challenges based on the previous days' discussions as well as drawing on expertise in their own domains and were tasked with distinguishing research priorities that could be characterized as short, medium, and long-term. The objective of defining such an agenda is to highlight and define areas of research that will benefit and support analytical efforts, as well as identify open research questions for the academic geographic information science (GIScience) community to address.

I.   RESEARCH CHALLENGES

Eight research challenges were identified at the workshop. These challenges include

- Data modeling for dynamic geographic domains
- Computational requirements for dynamics

- Visualization of geographic dynamics

- Spatiotemporal knowledge discovery

- Geographic dynamics over multiple granularities

- Spatiotemporal uncertainty and accuracy

- Dynamic social networks

- Feature extraction and analysis of images, video, and other unstructured dynamic information sources

The broad topic of each challenge area requires medium- to long-term support with respect to research and funding, as these all represent important research areas with many open questions. The subtopics described within each challenge area typically represent shorter term projects, often at the granularity of a master's or doctorate dissertation.

A.   Data Modeling for Dynamic Geographic Domains

To date, most data modeling efforts have been focused on static *snapshots* that reduce the profoundly dynamic nature of the real world to simpler, abstracted perspectives that are fixed or stationary in some way. In an effort to move beyond the snapshot view to a more realistic, dynamic view, a new generation of data models is required. Several key questions need to be considered in order to move toward these new data models; for example, what kinds of geographic phenomena are fundamentally dynamic, what are the core characteristics of dynamic data, and what makes geographic data dynamic?

Related to this challenge of designing next-generation data models are further investigations into topics that relate specifically to time. These topics include, for example, different kinds of temporal semantics and times (e.g., branching time, current time, or relative versus absolute time), data models that support cycles or cyclic events, as well as designing data models using event or process-based approaches.

Other data modeling research challenges exist where time and space together are primary considerations. Data models for moving entities (e.g., people, animals, vehicles, or other entities such as pollution clouds or protest marches) have been a topic of consideration by the GIScience community as well as the database and image processing communities. To support the data modeling of moving entities, topics including, for example, event modeling, spatiotemporal database design, and spatiotemporal query languages, require further attention.

Also under the umbrella of research needed for next-generation dynamic data modeling is the topic of dynamic ontologies. Ontologies have been constructed for many application domains, however, primarily based on static views of a domain. Extending research on ontologies such that ontologies support dynamic domains is another important research thread relating to data modeling. This includes, for example, constructing and sharing dynamic ontologies as well as extending any semantic classification system to handle dynamic data and applications.

B.   Computational Requirements for Dynamics

An important research challenge is to develop system architectures that are designed for dynamic applications and analysis. To date, information system architectures have been focused primarily on the needs of static application domains. This step forward involves developing techniques for dynamic data fusion, as well as techniques for conflation and integration of dynamic data. An interesting part of this challenge is to study the role of computational techniques from the gaming and simulation communities, where dynamics have been introduced to lend a more realistic sense to the game or simulation (e.g., urban dynamics are more realistically rendered). Another important topic is that spatiotemporal data sets can be very large and, therefore, new methods for spatiotemporal data storage, indexing, and multi-dimensional processing are needed. Although tools for time series analysis are being developed, in general, more computational tools and, thus, techniques for spatiotemporal statistical modeling are also needed. In general, intelligence problems are often problem-focused, and new techniques and tools that can define the road ahead are very important. This can mean developing computational tools that are more revolutionary than evolutionary.

C.   Visualization of Dynamic Processes

While visualization of dynamics is already a subject of major conferences (e.g., the IEEE Visualization Conference series and the IEEE Symposium on Visual Analytics Science and Technology) and centers (e.g., Charlotte Visualization Center and the National Visualization and Analytics Center), there are still so many open research questions that relate to this area that it merits high consideration on the *Research Challenges* list. During the workshop, presentations on the topic of visualization addressed the visual analysis of urban terrain, visual analysis of human activities, the nature and psychology of visual dynamics, and multi-dimensional urban visualizations, among others. Open research questions relate to the visualization of structured and unstructured dynamic data, cognitive aspects of dynamic visualizations (e.g., visualization and cognitive cartography), visualizations of spatiotemporal uncertainty, multi-scale dynamic visualizations, and visualizations of interacting processes.

D.   Dynamics over Multiple Granularities

Data modeling for dynamic geographic domains requires models that can treat multiple granularities of data and processes. Open research questions exist relating to generalization and multi-level, multi-resolution representations. Discussions at the workshop illustrated how important it is to consider multiple scales; for example, to consider local, statewide, and regional scales, as well as consider studies from the scale of an individual. Humanitarian relief must be understood at the scale of an individual, a city, as well as at the scale of a developing nation. Dynamics of urban areas *and* rural areas are important to understand. Other granularity-related topics relate to temporal modeling where there is a need to represent and perform analysis across different temporal granularities (e.g., yearly, daily, seasonally, etc.).

E.   Spatiotemporal Uncertainty and Accuracy

Extending research on uncertainty and accuracy to the case of spatiotemporal uncertainty and accuracy is also a key research challenge. Modeling, quantifying, and visualizing uncertainty of dynamics in geographic domains are all new and important topics for both researchers and analysts. This work aims to deliver a better understanding of, for example, the interdependence of spatial and temporal uncertainty, and includes studies of the qualitative representation of uncertainty, problems of missing data, and deception.

F.   Spatiotemporal Knowledge Discovery

An important research challenge relations to the topic of dynamic data exploration versus more structured, analytical methods typically applied to static data applications. Given a dynamic geographic domain, novel methods are needed for discovering and recognizing dynamic patterns, clusters, and hot spots. This extends work on data modeling, for example, going beyond modeling patterns of human activities (e.g., a daily activity pattern) to actually predicting or revealing previously unknown patterns of activities. New research areas involving, for example, cognitive-based knowledge discovery can be applied to the treatment of dynamics in geographic domains.

G.   Dynamic Social Networks

This research challenge focuses on the spatiotemporal aspects of human social networks where space and time are integrated tightly together. This area has links to many of the other challenge areas; for example, related topics are computational models for dynamic social networks, visualization of dynamic social networks, and modeling and representing uncertainty with respect to such a network. Social network analysis connects individuals and their social relationships, such as kinships, friendships, and other societal or religious associations. Research on dynamic social network analysis is needed to address how these social relationships change over time and how the changes may be predicted through investigation of spatiotemporal activities of individual or group interactions. Dynamics of physical and human environments can both promote and suppress certain social activities and social networking. A multi-level analysis of micro-, meso-, and macro-factors in space and time is necessary to comprehend the formation, strengthening, or weakening social networks as well as transforming members and functions of social networks. Research on event modeling and modeling moving entities can also be related to this challenging topic.

H.   Feature Extraction and Analysis of Images, Video, and Other Unstructured Information Sources

Video data collections capture dynamic information about a domain, which can include, for example, city streets, subways, or videos of crowds attending sporting events. These videos incorporate semantics associated with dynamic happenings or events (e.g., a subway train leaving the platform) and new feature extraction methods based on dynamics rather than static features are necessary. Even static imagery contains content relating to dynamics, and tools and techniques for dynamic feature extraction must be developed.

II.   RECOMMENDATIONS

Discussions among participants generated several recommendations to address some of the difficulties experienced, on the one hand, by analysts trying to work with dynamic data, and, on the other hand, by academic researchers trying to understand the most pressing geospatial problems for intelligence analysts.

A.   Develop Benchmark Dynamic Data Sets

The first of these recommendations is to generate a set of benchmark data collections and benchmark scenarios that can be used to gauge research results. The benchmark data set gives a common ground for researchers to explore their ideas on, for example, spatiotemporal uncertainty or perhaps develop a simulation. From the analysis side, making available some example situations and/or exercises that are unclassified for academics to study is an important aid in communicating what are some of the most pressing challenges for analysts.

B.   Build Strategies for Communication and Collaboration

A strong recommendation was made for increased opportunities for sustained communication and collaboration between analysts and academic researchers. Currently, some government agencies support visiting scholar programs where academics can come and spend a period of time with an agency, as well as supporting sending their employees to academic departments to get graduate degrees. More analysts may explore what possibilities exist for collaborating with researchers. Workshops and symposia also provide excellent opportunities for collaboration and communication as well as more extended forums such as summer institutes. Collaborative wikis designed to disseminate research results and histories of research efforts were also proposed as a means for analysts and academics to communicate and collaborate.

C. Build Strategies for Migrating from Research to Product Development or Application

This recommendation focuses on the need—particularly from the point of view of analysts—for transferring more top scientific ideas into usable software tools and techniques. As there are still few options available in commercial software for the representation and manipulation of time-varying geospatial data, greater efforts need to be made to transfer research findings into usable products that can be applied when reasoning about dynamics in geographic domains.

D. Develop Dynamic Geospatial Education Opportunities

Academics in fields relating to geographic information science are recommended to incorporate spatiotemporal topics into their curricula, producing a body of educated experts who are trained and experienced in working with dynamic data. Efforts such as the *Geographic Information Science and Technology Body of Knowledge* published by the University Consortium for Geographic Information Science (UCGIS), which sets the underpinning body of knowledge for those studying geographic information science, for example, could emphasize the important role of dynamics for geographic domains (Dibiase et al. 2006)

REFERENCES

Abrahams, A.D. 1984. Channel networks: A geomorphological perspective. *Water Resources Research* 20: 161-168.

Ahl, V. and T.F.H. Allen. 1996. *Hierarchy Theory: A Vision, Vocabulary, and Epistemology.* New York: Columbia University Press.

Allen, J. F. 1983. Maintaining knowledge about temporal intervals. *Commun. ACSM* 26(1): 832-843.

Andrienko, N., J. Hipolito, U. Kretchmer, G. Andrienko, H. Voss, and F. Bernardo. 2002. Testing the usability of interactive maps in commonGIS. *Cartography and Geographic Information Science* 29(4): 325-342.

Armstrong, M.P. 1988. Temporality in spatial databases. *Proceedings: GIS/LIS'88*, 2: 880-889.

Armstrong, M.P. 1991. Knowledge classification and organization. *Map generalization.* B.P. Buttenfield and R.B. McMaster eds., New York: Longman, Harlow/Wiley, 86-102.

Barabási, A.L. 2005. Network theory—the emergency of the creative enterprise. *Science* 308(5722): 639-641.

Barlett, M.S. 1956. Deterministic and stochastic models for recurrent epidemics. *Proceedings of the Third Berkeley Symposium on Mathematical Statistics and Probability*, 1954-1955, IV, 81-109. Berkeley and Los Angeles: University of California Press.

Batty, M. 2005. *Cities and Complexity.* Cambridge, MA: Massachusetts Institute of Technology.

Bennett, D.A. and W.Tang. 2006. Modelling adaptive, spatially aware, and mobile agents: Elk migration in Yellowstone. *International Journal of Geographical Information*, 20(9): 1039-1066.

Bithell, M. and W.D. Macmillan. 2007. Escape from the cell: Spatially explicit modelling with and without grids. *Ecological Modelling*, 200: 59-78.

Blaut, J.M. 1961. Space and processes. *The Professional Geographer: The Journal of the Association of American Geographers* 13(4): 1-7.

Bonacich, P. 2003. Cellular automata for the network researcher. *Journal of Mathematical Sociology* 27: 263-278.

Bone, C., S. Dragicevic and A. Roberts. 2006. A fuzzy-constrained cellular automata model of forest insect infestations. *Ecological Modelling* 192(1-2): 107-125.

Bowers, K.J., S.D. Johnson and K. Pease. 2004. Prospective hot-spotting, the future of crime mapping? *The British Journal of Criminology* 44(5): 641-658.

Box-Steffensmeier, J.M. and B.S. Jones. 2004. *Event History Modeling: A Guide for Social Scientists.* Cambridge, UK: Cambridge University Press.

Brogan, D.C., R.A. Metoyer, and J.K. Hodgins. 1998. Dynamically simulated characters in virtual environments. *Computer Graphics and Applications, IEEE* 18(5): 58-69.

Campos, J. and K. Hornsby. 2004. Temporal constraints between cyclic geographic events, *Proceedings of VI Brazilian Symposium on GeoInformatics, GeoInfo 2004,* Campos do Jordao, Brazil, 109-125.

Cañas, A.J., M. Carvalho, M. Arguedas, D.B. Leake, A. Maguitman, and T. Reichherzer. 2004a. Mining the web to suggest concepts during concept map construction, *Concept Maps: Theory, Methodology, Technology. Proceedings of the First International Conference on Concept Mapping*, J. Cañas, J.D. Novak, and F.M. González, eds., Pamplona, Spain : Universidad Pública de Navarr, 135-142.

Cañas, A.J., R. Carff, G. Hill, M. Carvalho, M. Arguedas, T.C. Eskridge, J. Lott, and R. Carvajal. 2004b. Googling from a concept map: Towards automatic concept-map-based query formation, *Concept Maps: Theory, Methodology, Technology. Proceedings of the First International Conference on Concept Mapping, A.J. Cañas, J.D. Novak, and F.M. González, eds. Pamplona, Spain: Universidad Pública de Navarra, 409-416.

Chen, Q., A. E. Mynett, and A. W. Minns. 2002. Application of cellular automata to modelling competitive growths of two underwater species *Chara aspera* and *Potamogeton pectinatus* in Lake Veluwe. *Ecological Modelling* 147(3): 253-265.

Chrisman, N.R. 1998. Beyond the snapshot: Changing the approach to change, error, and process. *Spatial and Temporal Reasoning in Geographic Information Systems.* M.J. Egenhofer and R. G. Golledge, eds. New York: Oxford University Press, 85-93.

Cliff, A. and P. Haggett. 2006. A swash-backwash model of the single epidemic wave. *Journal of Geographical Systems.* 8(3): 227-252.

Cohen, L. E. and M. Felson. 1979. Social change and crime rate trends: A routine activity approach. *American Sociological Review* 44: 588-608.

Cole, S. and K. Hornsby. 2005. Modeling noteworthy events in a geospatial domain, *Proceedings of the First International Conference on Geospatial Semantics*, GeoS 2005, Lecture Notes in Computer Science, 3799, Berlin: Springer Verlag, 78-92.

Davis, M.W. 1889. The rivers and valleys of Pennsylvania. *National Geographic Magazine* I: 183-253.

Dibiase, D., A.M. Maceachren, J.B. Krygier, and C. Reeves. 1992. Animation and the role of map design in scientific visualization. *Cartography & Geographic Information Systems* 19(4): 201-214.

Dibiase, D., M.N. DeMers, K.Kemp, A. Johnson, B. Plewe, and A Luck. 2006. *Geographic Information Science and Technology Body of Knowledge*—University Consortium for Geographic Information Science. Washington, DC: American Association for Geographers.

Dietzel, C. and K. Clarke. 2006. The effect of disaggregating land use categories in cellular automata during model calibration and forecasting. *Computers, Environment and Urban Systems* 30(1): 78-101.

Ducheyne, E.I., R.R. De Wulf, and B. De Baets. 2006. A spatial approach to forest management optimization: Linking GIS and multiple objective genetic algorithms. *International Journal of Geographical Information Science* 20(8): 917-928.

Dussutour, A., V. Fourcassie, D. Helbing, and J. L. Deneubourg. 2004. Letters to nature—Optimal traffic organization in ants under crowded conditions. *Nature* 428(6978): 70-72.

Egenhofer, M. and R. Franzosa. 1991. Point-set topological spatial relations. *International Journal of Geographical Information Systems* 5(2): 161-174.

Fabrikant, S.I., D.R. Montello, and D.M. Mark. 2006. The distance-similarity metaphor in region-display spatializations. *Computer Graphics and Applications, IEEE* 26(4): 34-44.

Favier, C., D. Schmit, C.D. M. Müller-Graf, B. Cazelles, N. Degallier, B. Mondet, and M.A. Dubois. 2005. Influence of spatial heterogeneity on an emerging infectious disease: The case of dengue epidemics. *Proceedings: Biological Sciences* 272(1568): 1171-1177.

Fonnesbeck, C. J. and M. J. Conroy. 2004. Application of integrated Bayesian modeling and Markov Chain Monte Carlo methods to the conservation of a harvested species. *Animal Biodiversity and Conservation* 27(1): 267-281.

Friedman, D., A. Shamir, Y. A. Feldman, and T. Dagan. 2004. Automated creation of movie summaries in interactive virtual environments. *IEEE Proceedings: Virtual Reality* 2004, 191-209.

Gadia, S. K. and J. H. Vaishnav. 1985. A query language for a homogeneous temporal database. *Proceedings of the ACM Symposium on Principles of Database Systems*, 51-56.

Gadia, S. K. and C. S. Yeung. 1988. A generalized model for a relational temporal database. *Proceedings of ACM SIGMOD International Conference on Management of Data*, 251-259.

Gahegan, M. 1999. Guest Editorial: What is geocomputation? *Transactions in GIS* 3(3): 203-206.

Gahegan, M., M. Wachowicz, M. Harrower and T. M. Rhyne. 2001. The integration of geographic visualization with knowledge discovery in databases and geocomputation. *Cartography and Geographic Information Science* 28(1): 29-44.

Gahegan, M., F. Hardisty, M. Takatsuka, M. Wheeler, I. Lee, and M. Gahegan. 2002. Introducing Geo VISTA Studio: An integrated suite of visualization and computational methods for exploration and knowledge construction in geography. *Computers, Environment and Urban Systems* 26(4): 267-292.

Galton, A. and M. Worboys. 2005. Processes and events in dynamic geo-networks. *Lecture Notes in Computer Science,* Berlin: Springer Verlag, 3799: 45-59.

Godsill, S.J. 2001. On the relationship between Markov Chain Monte Carlo Methods for model uncertainty. *Journal of Computational and Graphical Statistics: A joint publication of American Statistical Association,* Institute of Mathematical Statistics, Interface Foundation of North America. 10(2): 230-248.

Goodchild, M.F., M. Yuan, and T. J. Cova. 2007. Towards a general theory of geographic representation in GIS. *International Journal of Geographic Information Science* 21(3): 239-260.

Gordon, T. J., R. R. Sengupta, and D. A. Bennett. 2003. A simple agent model of an epidemic: Agent-based modelling environment for spatial decision support. *Technological Forecasting and Social Change* 70(5): 397-417.

Grenfell, B. T., O. N. BjØrnstad, and J. Kappey. 2001. Trevalling waves and spatial hierarchies in measles epidemics. *Nature* 414: 715-723.

Grenon, P. and B. Smith. 2004. SNAP and SPAN: Towards dynamic spatial ontology. *Spatial Cognition and Computation* 1: 69-104.

Guo, D., D. J. Peuquet, and M. Gahegan. 2003. ICEAGE: Interactive clustering and exploration of large and high-dimensional geodata: Is inductive machine learning just another wild goose (or might it lay the golden egg)? *GeoInformatica* 7(3): 229-253.

Guo, D., M. Gahegan, A. M. MacEachren, and B. Zhou. 2005. Multivariate analysis and geovisualization with an integrated geographic knowledge discovery approach. *Cartography and Geographic Information Science* 32(2): 113-132.

Guo, D., J. Chen, A. M. MacEachren, and K. Liao. 2006. A visualization system for space-time and multivariate patterns (VIS-STAMP). *Visualization and Computer Graphics, IEEE Transactions on* 12(6): 1461-1474.

Haario, H., M. Laine, M. Lehtinen, E. Saksman, and J. Tamminen. 2004. Markov Chain Monte Carlo methods for high dimensional inversion in remote sensing. *Journal of the Royal Statistical Society: Series B (Statistical Methodology)* 66(3): 591-607.

Hägerstrand, T. 1970. What about people in regional science? *Papers of the Regional Science Association* 24: 1-21.

Haklay, M., D. O'Sullivan, M. Thurstain-Goodwin, and T. Schelhorn. 2001. "So go downtown": Simulating pedestrian movement in town centres. *Environment and Planning B: Planning and Design* 28(3): 343-359.

Harrower, M. 2002. Visualizing change: Using cartographic animation to explore remotely sensed data. *Cartographic Perspectives* 39: 30-42.

Harrower, M., A. MacEachren, and A. L. Griffin. 2000. Developing a geographic visualization tool to support earth science learning. *Cartography and Geographic Information Science* 27(4): 279-293.

Harvey, A. and P. A. Macnab. 2000. Who's up? Global interpersonal temporal accessibility. *Information, Place and Cyberspace: Issues in Accessibility*. D. Janelle and D.C. Hodge, eds. Berlin: Springer Verlag, 147-170.

Helbing, D., P. Molnar, I.J. Farkas, and K. Bolay. 2001. Self-organizing pedestrian movement. *Environment and Planning B: Planning and Design* 28: 361-383.

Holawe, F. and R. Dutter. 1999. Geostatistical study of precipitation series in Austria: Time and space. *Journal of Hydrology* 219(1-2): 70-82.

Hornsby, K. and M. Egenhofer. 2000. Identity-based change: a foundation for spatio-temporal knowledge representation. *International Journal of Geographical Information Science* 14(3): 207-224.

Hornsby, K. and M. Egenhofer. 2002. Modeling moving objects over multiple granularities. *Annals of Mathematics and Artificial Intelligence* 36(1-2): 177-194.

Hsieh, J.W. and Y.S. Huang. 2002. Multiple-person tracking system for content analysis. *International Journal of Pattern Recognition and Artificial Intelligence* 16(4): 447-462.

Huang, B., R.L. Cheu, and Y. S. Liew. 2004. GIS and genetic algorithms for HAZMAT route planning with security considerations. *International Journal of Geographical Information Science* 18(8): 769-787.

Huntington, S. P. 1996. *The clash of civilizations and the remaking of world order*. New York: Simon and Schuster.

Janelle, D. 1995. Metropolitan expansion, telecommuting and transportation. *The Geography of Urban Transportation*. S. Hanson. New York: Guilford Press, 407-434.

Keim, D. A., C. Panse, M. Sips, and S. C. North. 2004. Visual data mining in large geospatial point sets. *Computer Graphics and Applications, IEEE* 24(5): 36-44.

Kerridge, J., J. Hine, and M. Wigan. 2001. Agent-based modelling of pedestrian movements: the questions that need to be asked and answered. *Environment and Planning B: Planning and Design* 28(3): 327-341.

Kohonen, T. 2001. *Self-Organizing Maps*. Berlin and New York: Springer Verlag.

Koua, E., A. Maceachren, and M. Kraak. 2006. Evaluating the usability of visualization methods in an exploratory geovisualization environment. *International Journal of Geographic Information Science, 20(1):425-448.*

Kraak, M.J. 2004. Visualization Viewpoints—Geovisualization for knowledge construction and decision support, geovisualization illustrated. *IEEE Computer Graphics and Applications* 24(1): 13-17.

Krzanowski, R. and J. Raper. 2001. *Spatial Evolutionary Modeling*. Oxford, UK: Oxford University Press.

Kumar, P., D. Gospodaric, and P. Bauer. 2007. Improved genetic algorithm inspired by biological evolution. *Soft Computing* 11(10): 923-941.

Kupfer, J.A. and J. R. Runkle. 2003. Edge-mediated effects on stand dynamic processes in forest interiors: A coupled field and simulation approach. *Oikos* 101(1): 135-146.

Kwan, M.P. 1999. Gender and individual access to urban opportunities: A study using space-time measures. *The Professional Geographer* 51(2): 211-227.

Kwan, M.P. 2001. Interactive geovisualization of activity-travel patterns using three-dimensional geographical information systems: A methodological exploration with a large data set. *Sage Urban Studies Abstracts* 2(1): 3-135.

Kwan, M.P. 2004. GIS methods in time-geographic research: Geocomputation and geovisualizationhot of human activity patterns. *Geografiska Annaler, Series B: Human Geography* 86(4): 267-280.

Kwan, M.P. and N. Schuurman. 2004. Special issue: Addressing the social implications of GIS in the technical realm, I: Public health data and GIS—introduction: Issues of privacy protection and analysis of public health data. *Cartographica* 39(2): 1-3.

Langran, G. and N. Chrisman, R. 1988. A framework for temporal geographic information. *Cartographica* 25(3): 1-14.

Laube, P. and S. Imfeld. 2003. Analyzing relative motion within groups of trackable moving point objects. *Proceedings: GIScience 2002*, M. Egenhofer and D. Mark, eds. Berlin: Springer Verlag, 132-144.

Laube, P., S. Imfeld, and R. Weibel. 2005. Discovering relative motion patterns in groups of moving point objects. *International Journal of Geographic Information Science* 19(6): 639-668.

Lee, Y. and D. W. Nelson. 2005. Viewing or visualizing, which concept map strategy works best on problem-solving performance? *British Journal of Educational Technology* 36(2): 193-203.

Levine, N. 2006. Crime mapping and the crimestat program. *Geographical Analysis* 38(1): 41-56.

Lewis, S. M. and A. E. Raftery. 1999. Bayesian analysis of event history models with unobserved heterogeneity via Markov Chain Monte

Carlo: Application to the explanation of fertility decline. *Sociological Methods & Research* 28(1): 35-60.

Li, X. and A. G. O. Yeh. 2005. Integration of genetic algorithms and GIS for optimal location search. *International Journal of Geographical Information Science* 19(5): 581-601.

Liu, Y. and S. R. Phinn. 2003. Modelling urban development with cellular automata incorporating fuzzy-set approaches. *Computers, Environment and Urban Systems* 27(6): 637-658.

Loreau, M., S. Naeem, P. Inchausti, J. Bengtsson, J. P. H. Grime, A.D. Hooper, M.A. Huston, D. Raffaelli, B. Schmid, and D. A. Wardle. 2001. Biodiversity and ecosystem functioning: Current knowledge and future challenges. *Science* 294(5543): 804-808.

Lucas, M. C. and E. Baras. 2000. Methods for studying spatial behaviour of freshwater fishes in the natural environment. *Fish and Fisheries* 1(4): 283-316.

MacEachren, A.M., M. Gahegan,, and W. Pike. 2004. Colloquium paper— Visualization for constructing and sharing geo-scientific concepts. *Proceedings of the National Academy of Sciences of the United States of America* 101(1): 5279-5286.

MacEachren, A. M., R. Murray, M. Gahegan, E. Hetzler, A. Robinson, S. Hopper, and S. Gardner. 2005. Visualizing geospatial information uncertainty: What we know and what we need to know. *Cartography and Geographic Information Science* 32(3): 139-160.

MacEachren, A. M., I. Brewer, M. Gahegan, S. D. Weaver, B. Yarnal, W. Pike and C. Yu. 2006. Building a geocollaboratory: Supporting human-environment regional observatory (HERO) collaborative science activities. *Computers, Environment and Urban Systems* 30(2): 201-225.

Maguire, D., M. Batty, and M. Goodchild. 2005. *GIS, Spatial Analysis, and Modeling*. Redlands, CA: ESRI Press.

Mark, D. and M. Egenhofer. 1994. Modeling spatial relations between lines and regions: Combining formal mathematical models and human subjects testing. *Cartography and Geographical Information Systems* 21(3): 195-212.

Mark, D. M. and M. J. Egenhofer. 1998. *Geospatial lifelines*. Dagstuhl Seminar Report No. 228.

Martin, Y. 2000. Modelling hillslope evolution: Linear and nonlinear transport relations. *Geomorphology* 34(1-2): 1-21.

Mass, D. S. and M. Yuan. 1999. Analyzing dispersal dynamics of golden-cheeked warblers (*Dendroica chrysoparia*) using remote sensing, global positioning system, and geographic information system technologies. *Applied Geographic Studies* 3(2): 77-95.

Matthews, K.B., S. Craw, K. Buchan, and A. R. Sibbald. 2006. Combining deliberative and computer-based methods for multi-objective land-use planning. *Agricultural Systems* 87(1): 18-37.

McIntosh, J. and M. Yuan. 2005a. Assessing similarity of geographic processes and events. *Transactions in GIS* 9(2): 223-245.

McIntosh, J. and M. Yuan. 2005b. A framework to enhance semantic flexibility for analysis of distributed phenomena. *International Journal of Geographic Information Science* 19(10): 999-1018.

Meyer, M. and E. Miller 2001. *Urban Transportation Planning*. New York: The McGraw-Hill Companies, Inc.

Milgram, S. 1967. The small world problem. *Psychology Today*, 60-67.

Miller, H. J. 2005. Necessary space-time conditions for human interaction. *Environment and Planning B: Planning and Design* 32(3): 381-401.

Millington, J.D.A., G.L.W. Perry, and B.D. Malamud. 2006. Models, data mechanisms: Quantifying wildfire regimes. *Fractal Analysis for Natural Hazard,* United Kingdom Series: Geological Society Special Publication, 261: 155-167.

Mills, H. H. and R. T. Mills. 2001. Evolution of undercut slopes on abandoned incised meanders in the Eastern Highland Rim of Tennessee, USA. *Geomorphology* 38(3-4): 317-336.

Mitasova, H., L. Mitas, C. Ratti, H. Ishii, J. Alonso, and R. S. Harmon. 2006. Real-time landscape model interaction using a tangible geospatial modeling environment. *Computer Graphics and Applications, IEEE* 26(4): 55-63.

Monmonier, M. 1992. Summary graphics for integrated visualization in dynamic cartography. *Cartography & Geographic Information Systems* 19(1): 23-36.

Moulin, B. 1997. Temporal contexts for discourse representation: An extension of the conceptual graph approach. *Applied Intelligence* 7(3): 227.

Nesbit, J. C. and O. O. Adesope. 2006. Learning with concept and knowledge maps: A meta-analysis. *Review of Educational Research* 76(3): 413-448.

Ngo, C.W., Y.F. Ma, and H.J. Zhang. 2005. Video summarization and scene detection by graph modeling. *IEEE Transactions on Circuits and Systems for Video Technology,* 15(2): 296-305.

Nobre, A.A., A.M. Schmidt, and H.F. Lopes. 2005. Spatio-temporal models for mapping the incidence of malaria in Pará. *Environmetrics* 16(3): 291-304.

Noldus, L.P.J.J., A.J. Spink and R.A.J. Tegelenbosch. 2002. Computerised video tracking, movement analysis and behaviour recognition in insects. *Computers and Electronics in Agriculture* 35(2): 201-228.

O'Neill, P.D., D.J. Balding, N.G. Becker, M. Eerola, and D. Mollison. 2000. Analyses of infectious disease data from household outbreaks by Markov Chain Monte Carlo methods. *Applied Statistics* 49(4): 517-543.

Ou, Y., H. Qian, X. Wu, and Y. Xu. 2005. Real-time surveillance based on human behavior analysis. *International Journal of Information Acquisition* 2(4): 353-365.

Pain, R., R. MacFarlane, K. Turner, and S. Gill. 2006. 'When, where, if, and but:' qualifying GIS and the effect of streetlighting on crime and fear. *Environment and Planning A* 38(11): 2055-2074.

Parker, D. C., P. Deadman, S. M. Manson, M. A. Janssen, and M. J. Hoffmann. 2003. Multi-agent systems for the simulation of land-use and land-cover change: A review. *Annals of the Association of American Geographers* 93(2): 314-337.

Peuquet, D. J. and N. Duan. 1995. An event-based spatiotemporal data model (ESTDM) for temporal analysis of geographical data. *International Journal of Geographical Information Systems* 9(1): 7-24.

Rana, S. and J. Dykes. 2003. A framework for augmenting the visualization of dynamic raster surface. *Information Visualization* 2: 126-139.

Randell, D.A., Z. Cui, and A.G. Cohn. 1992. A spatial logic based on regions and connection. *The 3rd International Conference on Knowledge Representation and Reasoning*, Burlington, MA: Morgan Kaufman, 165-176.

Rass, L. and J. Radcliffe. 2003. *Spatial Deterministic Epidemics*. Providence, RI: The American Mathematical Society.

Ratcliffe, J.H. and M. J. McCullagh. 1998. Aoristic crime analysis. *International Journal of Geographical Information Systems* 12(7): 751-764.

Ratcliffe, J. H. 2002. Damned if you don't, damned if you do: crime mapping and its implications in the real world. *Policing & Society* 12(3): 211-225.

Ratti, C., B. Piper, H. Ishii, D. Frenchman, and Y. Wang. 2004. Tangible User Interfaces (TUIs): A novel paradigm for GIS. *Transactions in GIS* 8(4): 407-421.

Reluga, T., J. Medlock, and A. Galvani. 2006. A model of spatial epidemic spread when individuals move within overlapping home ranges. *Bulletin of Mathematical Biology* 68(2): 401-416.

Rey, S. J. and M. V. Janikas. 2006. STARS: Space-time analysis of regional systems. *Geographical Analysis* 38(1): 67-86.

Rodriguez-Iturbe, I., E. Ijjasz-Vasquez, A. Marani, A. Rinaldo, R. Rigon, and R. L. Bras. 1992. Fractal structures as least energy patterns: The case of river networks. *Geophysical Research Letters* 19(9): 889-892.

Sadahiro, Y. 2001. Exploratory method for analyzing changes in polygon distributions. *Environment and Planning B: Planning and Design* 28(4): 595-609.

Sander, L.M., C. Simon, J. Koopman, C.P. Warren, and I.M. Sokolov. 2002. Percolation on heterogeneous networks as a model for epidemics. *Mathematical Biosciences* 180: 293-305.

Sauer, C. 1941. Forward to historical geography. *Annals of the Association of American Geographers* 31: 1-24.

Sauer, C. 1956. The education of a geographer. *Annals of the Association of American Geographers* 46: 287-299.

Scott, J. 2000. *Social Network Analysis: A Handbook*, Thousand Oaks, CA: Sage Publications Ltd.

Shreve, R. L. 1975. Probabilistic–topologic approach to drainage-basin morphology. *Geology* 3, 527-529.

Sinha, G. and D. M. Mark. 2005. Measuring similarity between geospatial lifelines in studies of environmental health. *Journal of Geographical Systems* 7(1): 115-136.

Sivic, J., F. Schaffalitzky, and A. Zisserman. 2006. Object level grouping for video shots. *International Journal of Computer Vision* 67(2): 189-210.

Slocum, T.A., S.C. Yoder, F.C. Kessler, and R.S. Sluter. 2000. Maptime: Software for exploring spatiotemporal data associated with point locations. *Cartographica*. 37: 15-32.

Smart, J. S. 1979. Deterministicism and randomness in fluvial geomorphology. *EOS, Transaction of American Geophysical Union* 60: 651-655.

Smith, J. R. 1999. VideoZoom spatio-temporal video browser. *IEEE Transactions on Multimedia,* 1(2): 157-171.

Snodgrass, R. and I. Ahn. 1986. Temporal databases. *IEEE Computer, September*, 35-42.

Sowa, J. F. 1984. *Conceptual Structures: Information Processing in Mind and Machines*. Readings, MA: Addison-Wesley.

Sowa, J. F. 2000. *Knowledge Representation: Logical, Philosophical, and Computational Foundations*. Pacific Grove, CA: Brooks Cole Publishing Co.

Stewart Hornsby, K. and S. Cole. 2007. Modeling moving geospatial objects from an event-based perspective. *Transactions in GIS*, 555-573.

Thomas, J. J. and K. A. Cook, eds. 2005. *Illuminating the Path: The Research and Development Agenda for Visual Analytics*, Piscataway, NJ: IEEE Press.

Thomas, J. J. and K. A. Cook. 2006. A visual analytics agenda. *Computer Graphics and Applications, IEEE,* 26(1): 10-13.

Tobler, W. R. 1970. A computer movie simulating urban growth in the Detroit region. *Economic Geography* 46: 234-240.

Tomassini, L., P. Reichert, R. Knutti, T.F. Stocker, and M.E. Borsuk. 2007. Robust Bayesian uncertainty analysis of climate system properties using Markov Chain Monte Carlo Methods. *Journal of Climate* 20(7): 1239-1254.

Torrens, P.M. and I. Benenson. 2005. Geographic automata systems. *International Journal of Geographical Information Science* 19(4): 385-412.

Van Dijk, S., D. Thierens, and M. De Berg. 2002. Using genetic algorithms for solving hard problems in GIS. *GeoInformatica* 6(4): 381-413.

Veeraraghavan, A., A. K. Roy-Chowdhury, and R. Chellappa. 2005. Face and gesture–matching shape sequences in video with applications in human movement analysis. *IEEE Transactions on Pattern Analysis and Machine Intelligence* 27(12): 1896-1910.

Verhofen, M. 2005. Markov Chain Monte Carlo Methods in financial econometrics. *Financial Markets and Portfolio Management* 19(4): 397-405.

Wachowicz, M. 1999. *Object-Oriented Design for Temporal GIS*. Bristol, PA: Taylor & Francis.

Warren, W. 2005. Hierarchy theory in sociology, ecology, and resource management: A conceptual model for natural resource or environmental sociology and socioecological systems. *Society & Natural Resources* 18(5): 447-466.

Weisburd, D. and C. Lum. 2005. The diffusion of computerized crime mapping in policing: Linking research and practice. *Police Practice and Research* 6(5): 419-434.

Wikle, C. K., L. M. Berliner, and N. Cressie. 1998. Hierarchical Bayesian space-time models. *Environmental and Ecological Statistics* 5(2): 117-154.

Woldernberg, M. J. 1969. Spatial order in fluvial systems—Horton's laws derived from mixed hexagonal hierarchies of drainage basin areas. *Geological Society of America Bulletin* 80: 97-112.

Wong, P. C. and J. Thomas. 2004. Visual analytics. *Computer Graphics and Applications, IEEE* 24(5): 20-21.

Worboys, M. 1994. A unified model of spatial and temporal information. *Computer Journal* 37(1): 26-34.

Worboys, M. 2005. Event-oriented approaches to geographic phenomena. *International Journal of Geographical Information Science* 19(1): 1-28.

Worboys, M. and K. Hornsby. 2004. From objects to events: GEM, the geospatial event model. *Proceedings: Geographic Information Science 2004*, Adelphi, MD: Springer Verlag, 327-344.

Wu, Q. 2000. The application of genetic algorithm in GIS network analysis. *International Archives of Photogrammetry and Remote Sensing*, 44(3): 1184-1191.

Wu, J. and J.L. David. 2002. A spatially explicit hierarchical approach to modeling complex ecological systems: Theory and applications. *Ecological Modelling* 153(1-2): 7-26.

Yuan, M. 1996. Modeling semantical, temporal, and spatial information in geographic information systems. *Geographic Information Research: Bridging the Atlantic*. M. Craglia and H. Couclelis, eds. London: Taylor & Francis, 334-347.

Yuan, M. 1999. Representing geographic information to enhance GIS support for complex spatiotemporal queries. *Transactions in GIS* 3(2): 137-160.

Yuan, M. 2001a. Geographic data structure. In *Geospatial Technology Manual*. J. D. Bossel, J. R. Jensen, R. B. McMaster, and C. Rizos, eds. Boca Raton, FL: Taylor & Francis.

Yuan, M. 2001b. Representing complex geographic phenomena with both object- and field-like properties. *Cartography and Geographic Information Science* 28(2): 83-96.

Zyda, M. 2005. From visual simulation to virtual reality to games. *Computer* 38(9): 25-32.

INDEX